9

Band 19 · September 1983

isotopenpraxis

AKADEMIE-VERLAG

ISSN 0021-1915 Isotopenpraxis, Berlin 19 (1983) 9, 293-328

EVP 9,- M

An unsere Autoren

Die Zeitschrift veröffentlicht Beiträge zu folgenden Themen: Kontrolle und Analyse radioaktiver und stabiler Isotope, Anwendung radioaktiver und stabiler Isotope in allen Gebieten der Forschung und Technik, der Industrie, Landwirtschaft und Medizin, Herstellung radioaktiver und stabiler Isotope sowie von markierten Verbindungen, kernelektronische Geräte für die Isotopenanwendung (Dickenmesser, Dichtemesser u. ä.) Detektoren für isotopentechnische Arbeiten und den Strahlenschutz in Isotopenlaboratorien, Meßtechnik und Strahlenschutz, Dosimetrie in Isotopenlaboratorien, kerntechnische Geräte für die Isotopenanwendung (Beton, Bestrahlungsanlagen u. ä.), Strahlenchemie, Radiochemie, ökonomische und perspektivische Fragen, Fragen der Weiterbildung.

1. Aufnahme von Manuskripten

Zur Veröffentlichung angenommen werden nur druckreife, noch nicht anderweitig publizierte Beiträge in deutscher, englischer oder russischer Sprache zum Themenbereich der Zeitschrift und zwar

Originalarbeiten
Kurze Mitteilungen
Aus der Praxis
Zusammenfassende Berichte
Tagungsberichte
Informationen.

Sind der Beitrag oder Teile daraus bereits veröffentlicht oder ist eine anderweitige Veröffentlichung vorgesehen, so muß dies der Redaktion mit genauer Angabe der Zweitveröffentlichung mitgeteilt werden. Mit der Annahme des Manuskriptes behält sich der Verlag das Recht der Vervielfältigung, Verbreitung und Übersetzung innerhalb der gesetzlichen Schutzfrist vor. Auszugsweiser Nachdruck oder photomechanische u. a. Wiedergabe sind an eine schriftliche Genehmigung des Verlages gebunden. Manuskripte, Zeichnungen und Photos von veröffentlichten Arbeiten werden nur auf den besonderen und bis zum Zeitpunkt der Veröffentlichung mitgeteilten Wunsch des Autors zurückgesandt.
Bei der unverlangten Einreichung von Manuskripten wird empfohlen, eine Befürwortung eines der Herausgeber oder Redaktionsausschußmitglieder beizufügen.
Manuskripte sind einzureichen bei der Redaktion, 1080 Berlin, Leipziger Str. 3–4

2. Form der Manuskripte

Die Autoren werden gebeten:
2 Manuskripte, eines davon in Original-Maschinenschrift, einseitig und mit 2zeiligem Abstand einzureichen; gut leserliche Photokopien sind zulässig; auf dem 1. Blatt (unter Weglassung der Bezeichnung)
den Titel der Arbeit
Vorname-Initialen und Namen sämtlicher Autoren
gegebenenfalls die Institution, in der sie tätig sind
eine kurze Zusammenfassung der Arbeit von höchstens 15 Zeilen
Keywords in alphabetischer Reihenfolge und
als Fußnote die Anschrift der Institution anzugeben;
wenn möglich, auf gesonderten Blättern Namen, Titel der Arbeit und Zusammenfassung in den beiden anderen obengenannten Sprachen 2fach beizufügen;
Literaturhinweise (fortlaufend numeriert), Fußnoten, Tabellen, Abbildungen (in 2–3facher Veröffentlichungsgröße), Abbildungsunterschriften (getrennt von den Abbildungen mit Zeichenerklärungen, Quellenangaben, falls erforderlich, und gegebenenfalls mit einem Vermerk für besondere Wünsche bezüglich der Abbildungsgröße), private und Dienstanschrift der Verfasser mit akademischen Titeln und ausgeschriebenen Vornamen auf besonderen Blättern ans Ende des Manuskriptes zu stellen;
Wünsche für den Kolumnentitel (Seitenüberschrift) bekanntzugeben;
Absätze durch Einzüge von mindestens 3 Anschlägen kenntlich zu machen;
die Maßeinheiten des SI-Systems ("Systeme International") zu verwenden;
Korrekturen im Manuskript unmittelbar an der zu korrigierenden Stelle anzubringen;
Formeln deutlich lesbar und möglichst groß einzutragen;
Farbabbildungen und nachträgliche Änderungen oder Ergänzungen in Abbildungen zu vermeiden;
hervorzuhebende Textstellen zu unterstreichen, und zwar halbfetten Satz mit einer geraden schwarzen Linie, kursiven Satz (Schrägschrift) mit einer blauen Wellenlinie (alle Eigennamen). Sperrungen gestrichelt;
besondere Buchstaben durch farbige Unterstreichung zu kennzeichnen, und zwar Formelzeichen in

Griechisch-Kursiv (α, β, ...) rot,
Kursiv (A, B, ...) blau
halbfetter Kursiv ($\boldsymbol{A}$, $\boldsymbol{B}$, ...) 2 Linien blau (Vektoren),
halbfetter Grotesk (**A**, **B**, ...) 2 Linien violett (Tensoren);

bei Literaturangaben auf Vollständigkeit zu achten. In nachstehender Reihenfolge müssen angegeben sein:

a) bei *Zeitschriften*: Vorname-Initialen und Namen sämtlicher Autoren, Titel der Zeitschrift (nach „Periodica Chimica" oder möglichst wenig gekürzt), Bandnummer (unterstrichen), Jahreszahl (in Klammern), Heftnummer (nur bei Zeitschriften, die jedes Heft mit S. 1 beginnen) sowie die Seitenzahl, auf der die zitierte Arbeit beginnt.

b) Bei *Büchern*: Initialen der Vornamen und Name des Autors (Herausgeber werden durch den Zusatz „ed." gekennzeichnet), voll ausgeschriebener Titel des Buches in der Originalsprache, evtl. Bandnummer, voll ausgeschriebener Name des Verlages, Nummer der Auflage (außer der ersten), Erscheinungsort(e), Erscheinungsjahr (hier nicht in Klammern setzen!) und evtl. Kapitelnummer, Seitenzahl u. ä.

c) bei *Sammelbänden, Tagungsberichten*: Es soll zusätzlich der Titel des zitierten Beitrags angegeben werden. Die Seitenzahl muß unbedingt angegeben werden. Aus dem Zitat muß hervorgehen, ob es sich um einen Tagungssammelband oder um einen noch nicht veröffentlichten Vortrag handelt. Zwischen Tagungsort und Erscheinungsort des entsprechenden Tagungsberichts ist zu unterscheiden.

3. Korrekturen

Der Autor erhält zwei Umbruchabzüge, von denen einer möglichst schnell korrigiert an die Redaktion zurückgereicht werden soll. Auf Wunsch des Autors kann die Korrektur auch von der Redaktion gelesen werden.

Durch nachträgliche Manuskriptänderungen verursachte Korrekturkosten gehen, sofern sie 10% der Satzkosten übersteigen, zu Lasten des Autors.

Der Autor erhält 50 Sonderdrucke kostenlos. Weitere Sonderdrucke können bis zur Rückgabe der Korrekturabzüge auf Kosten des Autors bei der Redaktion bestellt werden.

Bezugsmöglichkeiten

Bestellungen sind zu richten

– **in der DDR** an den Postzeitungsvertrieb unter Angabe der Kundennummer des Bestellenden oder an den AKADEMIE-VERLAG, DDR-1086 Berlin, Leipziger Straße 3–4, PF-Nr. 1233
– **im sozialistischen Ausland** an eine Buchhandlung für fremdsprachige Literatur oder an den zuständigen Postzeitungsvertrieb
– **in der BRD und Berlin(West)** an eine Buchhandlung oder an die Auslieferungsstelle KUNST UND WISSEN, Erich Bieber OHG, Wilhelmstraße 4–6, D-7000 Stuttgart 1
– **in den übrigen westeuropäischen Ländern** an eine Buchhandlung oder an die Auslieferungsstelle KUNST UND WISSEN, Erich Bieber GmbH, Dufourstraße 51, CH-8008 Zürich
– **im übrigen Ausland** an den Internationalen Buch- und Zeitschriftenhandel; den Buchexport, Volkseigener Außenhandelsbetrieb der Deutschen Demokratischen Republik, DDR-7010 Leipzig, Postfach 160; oder an den AKADEMIE-VERLAG, DDR-1086 Berlin, Leipziger Straße 3–4

Zeitschrift „Isotopenpraxis"

Herausgeber: Im Auftrag des Akademie-Verlages herausgegeben von Prof. Dr. Justus Mühlenpfordt
Verlag: Akademie-Verlag, DDR-1086 Berlin, Leipziger Straße 3–4, PF-Nr. 1233; Fernruf: 2236229; Telex-Nr.: 114420;
Bank: Staatsbank der DDR, Berlin, Kto.-Nr.: 6836-26-20712
Verantwortlich für den Inhalt des Heftes: Prof. Dr. H. Hübner, Leipzig
Anschrift der Redaktion (Fernruf Frau H. Schier: 6351177) der Zeitschrift Isotopenpraxis: Akademie-Verlag, DDR-1086 Berlin, Leipziger Str. 3–4, PF-Nr. 1233
Veröffentlicht unter der Lizenznummer 1527 des Presseamtes beim Vorsitzenden des Ministerrates der Deutschen Demokratischen Republik
Gesamtherstellung: VEB Druckerei „Thomas Müntzer", DDR-5820 Bad Langensalza
Erscheinungsweise: Die Zeitschrift „Isotopenpraxis" erscheint monatlich. Die 12 Hefte eines Jahrganges bilden einen Band. Bezugspreis je Band 180,– M zuzüglich Versandspesen (Preis für die DDR 108,– M) Preis je Heft 15,– (Preis für die DDR 9,– M)
Bestellnummer dieses Heftes: 1079/19/9

AN (EDV) 12831

ISOTOPENPRAXIS

Band 19 (1983) Heft 9, S. 293—328

Redaktion und Verlag: Akademie-Verlag, Redaktion Isotopenpraxis, 1086 Berlin, Leipziger Straße 3—4

Inhaltsverzeichnis

Originalarbeiten

Contents

Articles

Gemische von Polyäthylen geringer Dichte und feuerhemmenden Zusatzstoffen wurden bei einer Umgebungstemperatur von ca. 20 °C mit einer ^{60}Co-Gammaquelle und mit 13 MeV-Elektronen eines Linearbeschleunigers bestrahlt. Unmittelbar nach der Bestrahlung und nach einer Lagerung bei Zimmertemperatur wurden ESR-Messungen durchgeführt. Es wurde festgestellt, daß die Dominanz von Radikalen in der Probe von den Zusatzstoffen, der Bestrahlungsdosis und der Lagerungszeit abhängt. Bei steigender Dosis sinkt die Gesamtkonzentration von Radikalen in mit Elektronen bestrahlten Proben. Die Wechselwirkung zwischen der Fähigkeit zur Bildung von Allylradikalen und der Vernetzungswirkung wird anhand der unterschiedlichen Beweglichkeit der Radikale in Gegenwart und bei Fehlen von Zusatzstoffen erklärt.

H. Wagler

Substöchiometrische Bestimmung von Pd-II-Spuren mit Dithizon und Probleme der Bestimmung (*in Englisch*)

Isotopenpraxis 19 (1983) H. 9, S. 307 bis 309
2 Abb., 1 Tab., 7 Lit.

Dithizon ist ein Reagens, das häufig zur substöchiometrischen Bestimmung von Spuren an Metallen verwendet wird. Im Falle des Pd treten bei der Extraktion allerdings einige Probleme auf, da das Dithizon in der organischen Phase in zwei tautomeren Formen auftreten kann. Man nimmt gegenwärtig an, daß nur die Thiol-Form des Dithizons mit den Metall-Ionen reagiert. Durch Untersuchungen mit Hilfe des ^{103}Pd kann aber gezeigt werden, daß der Extraktionsmechanismus mit der Thion-Form eine bessere Interpretation von auftretenden Fehlerquellen zuläßt. Es werden einige Schlußfolgerungen daraus gezogen.

B. Heinrich, G. Musiol, U. Richter

Schnelle Elementsgehaltsbestimmung durch unelastische Neutronenstreuung und die Methode der zeitlich korrelierten assoziierten Teilchen (*in Englisch*)

Isotopenpraxis 19 (1983) H. 9, S. 309 bis 312
7 Abb., – Tab., 11 Lit.

Es wird das Meßprogramm der TU Dresden, Sektion Physik zur Elementgehaltsbestimmung beschrieben, das die γ-Strahlung nach der unelastischen Streuung von Neutronen einer Energie von 14 MeV nutzt. Zur Reduzierung der intensiven Untergrundstrahlung wird die Methode der zeitlich korrelierten assoziierten Teilchen (MEZKAT) verwendet. Erste experimentelle Ergebnisse von Kohlenstoff-, Silizium- und Sauerstoffbestimmungen an Gemischen aus Graphit und Sand werden vorgestellt. Es wird die Nutzung dieser Methode zur Bestimmung des Heizwertes von Braunkohle diskutiert.

N. S. Batschwarow, I. D. Wankow, L. P. Dimitrov, Ch. D. Shukow, P. A. Pawlow

Das Betareflektometer SR-77 (*in Englisch*)

Isotopenpraxis 19 (1983) H. 9, S. 312 bis 314
3 Abb., – Tab., – Lit.

Ein Betareflektometer zur quantitativen Bestimmung der Konzentration schwerer und halbschwerer Elemente wird beschrieben. Ein besonderes Merkmal des Geräts besteht in der direkten Ablesung der prozentualen Konzentration am Display-Gerät.

Kurze Originalmitteilungen

M. Kießling, B. Schossig

Zum thermischen Verhalten von Uranhydriden, die sich in der isotopen Zusammensetzung der Wasserstoffe unterscheiden

Isotopenpraxis 19 (1983) H. 9, S. 315 bis 317
– Abb., 3 Tab., 5 Lit.

Informationen

Isotopenpraxis 19 (1983) H. 9, S. 317 bis 320

Buchbesprechungen

Isotopenpraxis 19 (1983) H. 9, S. 320 bis 328

The blends of low density polyethylene and additives of flame retardants type were irradiated in a ^{60}Co gamma source and with 13 MeV electrons from linac, at temperature of environment about 20 °C. Esr measurements were performed just after irradiation and after storage at room temperature.
It was stated that dominated radical present in the sample depends on the additives, dose and storage time. With increasing dose the overall concentration of radicals in samples irradiated with electrons decrease. The correlation between ability to allyl radicals formation and efficiency of crosslinking is explained by differences of radicals mobility in the presence and absence of additives.

H. Wagler

Substoichiometric Determination of Pd-II-Traces with Dithizone and Problems of Determination (*in English*)

Isotopenpraxis 19 (1983) No. 9, p. 307 to 309
2 Fig., 1 Tab., 7 Ref.

Pd-traces determination by radioactive isotope dilution analysis based on solvent extraction of metal chelates – Dithizone is a reagent frequently used for the substoichiometric separation of trace amounts of metals. In the case of substoichiometric Pd-extraction there are some problems in the reproducibility of determination, because in solutions of dithizone in organic solvent exist two tautomeric species. At present it is supposed in the literature that only the thiol form of dithizone reacts with the metal ions. Extensive studies of reaction between Pd(II) and dithizone using ^{103}Pd suggest on extraction mechanism with the thione form. From this fact follows the general problems and the peculiarity of Pd-determination.

B. Heinrich, G. Musiol, U. Richter

Fast Analysis of Element Contents by Inelastic Scattering of Neutrons and the Method of Time Correlated Associated Particles (*in English*)

Isotopenpraxis 19 (1983) No. 9, p. 309 to 312
7 Fig., – Tab., 11 Ref.

The element content measurement programme of the Dresden University of Technology, Department of Physics, using the γ-radiation after inelastic scattering of 14 MeV neutrons is described. To suppress the intense background the Time Correlated Associated Particle Method (TCAPM) is used. First experimental results of carbon, silicon and oxygen content measurements with mixtures of graphite and sand are given. The use of this method for determining the calorific value of brown coal is discussed.

N. S. Batchvarov, I. D. Vankov, L. P. Dimitrov, Ch. D. Shukov, P. A. Pavlov

Betareflectometer SR-77 (*in English*)

Isotopenpraxis 19 (1983) No. 9, p. 312 to 314
3 Fig., – Tab., – Ref.

A betareflectometer for quantitative determination of heavy and medium heavy elements concentration is described. Important feature of the instrument is the direct read-out of the concentration in percents from the display.

Letters ot the Editor

M. Kießling, B. Schossig

On the Thermal Behaviour of Uranium Hydrides Having Different Isotopic Composition of Hydrogenes

Isotopenpraxis 19 (1983) No. 9, p. 315 to 317
– Fig., 3 Tab., 5 Ref.

Information

Isotopenpraxis 19 (1983) No. 9, p. 317 to 320

New Books

Isotopenpraxis 19 (1983) No. 9, p. 320 to 328

Isotopenpraxis 19 (1983) 9, S. 293 – 297

Direct Chemical Information from Special Radiotracers as well as from Outer X-Ray Excitation*)

G. Brunner

(Academy of Sciences of the GDR, Central Institute of Isotope and Radiation Research, Leipzig)[1])

Radionuclides which exhibit a larger interaction between nucleus and electron shell during their radioactive decay (especially electron capture and internal conversion) are, in principle, sensitive to their chemical surroundings. Consequently, they can be used to yield chemical information after an appropriate labelling procedure. The information can be extracted from the decay rate variation as well as from their X-ray emission as a response to an inner-shell vacancy creation during nuclear deexcitation. As an example of the former method the relationship between the decay rate of ^{99m}Tc and the chemical structures of pertechnetate and bis(meso-dimercapto-succinato)oxotechnetate(V) is considered.

Lower expectations with respect to measuring techniques are associated with the observation of X-ray emission. For a number of elements (especially in the 3d transition row) the K_α/K_β X-ray intensity ratio was found to be an indicator of the nearest neighbour atoms. Equivalent information can be drawn from outer X-ray excitation, too.

Keywords

chemical bonds; chemical state; decay; electron capture decay; half-life; inner-shell ionization; internal conversion; neutrinos; recoils; technetium 99; X-ray emission analysis

Introduction

Of course, in general it is nothing special to obtain chemical information from suitable radiotracer experiments, and the idea of getting analytical-chemical knowledge by radioactive indicators is nearly as old the scientific and industrial application of radioactivity itself. For instance the famous Ra indicator application by *Hahn* and *Straßmann* ensured the Ba identity as a reaction product after neutron-uranium interaction a difficult chemical surrounding.

But what is unavoidable in such experiments? To add further information in order to get a chemical answer in all cases! This may be an information on a local concentration resp. distribution of the tracer, for instance an answer to the question: *where* has the tracer gone and where is it now concentrated or distributed, or: in which fraction (as a function of *time*!) does the tracer come out? and so on.

Quite another quality of chemical information, that is in the sense of this lecture: real and direct chemical information could be obtained if radiation parameters like energy, intensity, spatial distribution of emission ... were dependent on the ligands of the emitting nuclide. To fulfil such a requirement an essential interaction between the chemically sensitive electron shell and the emitting nucleus of the radioactive nuclide must take place, an interaction including nuclear and thus radiative properties, that is the nucleus is needed to be more than a stiff potential centre emitting a simple "yes/no"-signal. But we do have possibilities in this respect, namely the useful hyperfine effects known as the *Mößbauer*effect, the distortion of angular radiation distributions and the influence on the nuclide half-lifes.

Whereas the *Mößbauer*-effect has been introduced as a powerful tool in analytics for a long time the distorted angular distributions are perhaps less meaningful for practical applications but shurley too badly known. The chemical variation of half-lifes is by far the worst known of these three hyperfine interaction fields. Small wonder! To measure a half-life at first means to measure a radiation intensity or — in other terms — a relative activity, and the standard errors in that field usually amount to as much as one per cent, but the "effects" typically lie in the 10^{-3} to 10^{4} region! This has been in the centre of our efforts for some years: to improve the accuracy of relative activity measurements. By using a most precise mechanical sample changing device with a rotating sample disc as well as a mechanical blocking and opening of radiation paths from several locally fixed radiation sources an error of the relative integral and differential measurements e.g. of the ^{99m}Tc-radiation and/or the γ- and X-radiation of ^{125}I of not more than 10^{-4} was reached. Besides the precise mechanics, advanced data processing and evaluation guaranteed this performance, whereas the usual electronics was used.

The first part of my lecture will point out that a change in half-life of ^{99m}Tc leads to some insights into Tc complexes which are of interest in nuclear medicine.

The second part is directed to the fact that the X-ray emission after an electron capture process (EC) depends in its relative intensity on the ligands of the atom under decay. This is in fact no hyperfine effect, but it demonstrates how to use the outer shell/inner shell interaction for obtaining direct chemical information by means of the radionuclide technique besides the well known "chemical shifts" of X-line sites in high-resolution X-ray spectroscopy.

Certainly, that type of information may also be obtained by other kinds of X-ray excitation, but there seems to be a specific aspect of the electron capture mode of excitation which leads to a greater chemical sensitivity in the per cent region or even higher, and which favours EC nuclides, especially of the first row of transition elements, like ^{49}V, ^{51}Cr, ^{54}Mn, ^{55}Fe, ^{57}Co, ^{56}Ni and ^{59}Ni and finally ^{64}Cu which can be regarded as a handful of radiotracers from a new point

*) Presented at the 2nd Working Meeting "Radioisotope Application and Radiation Processing in Industry", 28st September to 1th October 1982, Leipzig

[1]) Address: DDR-7050 Leipzig, Permoserstr. 15

of view as they are leading to direct and easily obtainable knowledge of their ligands.

1. The Δλ-Task

Our main effort was directed to ^{99m}Tc, because this has been a well known case for a chemically sensitive decay for some time. The low-energy transition of E3-type is highly converted, and the K and L shells are excluded from the conversion because they have a higher electron binding energy than the transition energy exhibits.

^{99m}Tc is of great interest in view of its relevance to the field of nuclear medicine: it is widely used in the nonsurgical diagnostics of internal organs because of its ready availability from the so-called Tc-generator, and its favourable γ-emission of ≈140 keV is commonly used for radioactive imaging of internal organs.

But obviously the chemists have some preparational problems with the Tc complexes.

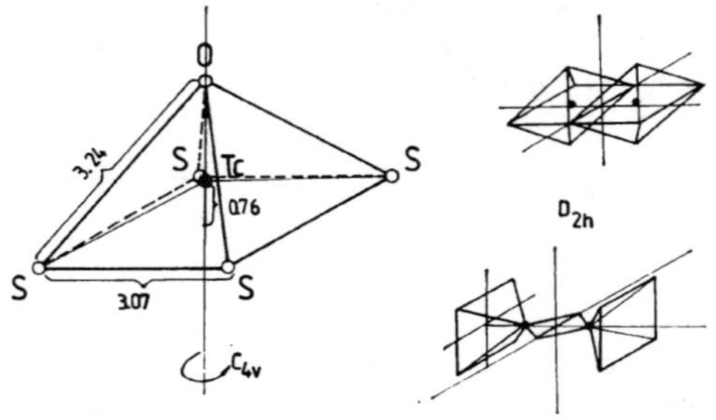

Fig. 1. Tc complexes considered
left: bis(dithiolato)oxotechnetate(V) complex ($TcOS_4$) – right: dimeric models of Tc(IV)

For instance, there seems to be a tendency in oxygen coordinated Tc(IV) with a coordination number of six in a normally octahedral configuration to form dimeric structures (Fig. 1, right hand side). Such problems are normally answered by optical absorption spectrometry if it is possible to offer a theoretically or, otherwise, experimentally founded alternative for characteristic energy levels of the models in question. And they have been answered in this way by a fruitful cooperation between our working group and the respective Rossendorf colleagues.

But: It is not easy to transfer such results of highly concentrated media to the carrier-free solutions of diagnostic imaging agents.

And this was our problem: Can it be assured that such high-concentration results are transferable to the carrier-free region?

Anticipating the final answer: yes, it can! We confirmed it for a Tc complex which is called bis(dithiolato)oxotechnetate(V) complex, $(TcOS_4)^-$, of a square-pyramidal structure with the sulfur atoms forming the basal plane and the oxygen atom the apex (Fig. 1, left hand side).

The basic idea is as follows: We measure the absorption bands in a sufficiently high concentration as well as the precise decay rate in the carrier-free preparation and also calculate both the optical absorption energies and the decay rate with respect to a well known standard like Tc-pertechnetate. If the theoretical results agree with both the experiments we have a high degree of certainty that the concentration of the preparation under consideration is of no influence.

Before I present the results, let me tell just a few words about the calculations. The so-called SCF-Xα-SW method was used. It is a fast working formalism for dealing with multicentre problems by splitting them into several single centre treatments on the Slater approximation level. The problem is, to reasonably connect the individual single potentials of a spherical kind with each other. This is performed by introducing an interatomic, resp. interspherical region of a constant and mean potential. This is the most cirtical point, and therefore this intermediate region should be kept at a minimum in its local extension.

The individual energetic terms, expressed by the so-called irreducible representations considering the complex symmetry, immediately result from the Xα-treatment. Localized at the Tc atom a dipole transition of moderate intensity between the 6e state at −5·81 eV and the 4e state at −8·71 eV was obtained. This leads to a transition energy equal to the difference, that is 2·9 eV. It is in a rather good agreement with some experimental absorption wavelengths $\lambda_{max} \approx 390 \ldots 415$ nm (3·0 ... 3·2 eV).

As far as the chemical action on the measurable Internal Conversion (IC) is concerned, it was not necessary to calculate it explicitely in the usual way from the radial parts of the potential waves for the different subshells, but instead a rescaling of the subshell conversion coefficients in terms of their electron densities at the nucleus was performed, referring to the respective density values of the free Tc atom in its equivalent Xα description.

Apart from the subshell fractions one obtains a relative (to the TcO_4^- ion) *decrease of the IC probability of 2·36 p.m.* in a fairly good agreement with the experimental value of

$$\Delta\lambda/\lambda = 2{\cdot}04 \pm 0{\cdot}10 \text{ p.m.}$$

And this is the state of our art: We learned from this that our Δλ-result of the carrier-free preparation corresponds to the optical absorption in a dense solution, and we think that it is a most valuable direct chemical information drawn from the ^{99m}Tc tracer by the Δλ technique which is really unique for this kind of question.

2. EC Nuclide Tracers

Now let me start with the second part of my talk: direct chemical information from EC nuclides.

For a lot of years the so-called chemical shifts have been a well known phenomenon in X-ray physics and its analytical applications. They rest upon the fact that even inner shell energies of atoms depend on ligands. That means, also X-ray line sites are influenced by ligands. But because X-ray energies represent *differences* of inner-shell energy states, the chemical shifts of X-ray lines are only of a small amount (eV), and this demands high-resolution spectrometers. Thus the general investigation of chemical effects on atomic energy levels shifted towards electron spectroscopy methods (ESCA), because they indicate the single level effects instead of their differences. But even newer is the aspect of line intensities as they are influenced by ligands, because it can be realized with a sufficient accuracy by simple multichannel analyzing of X-radiation. The respective requirements can be met best by the well-known energy-dispersive spectrometers including semiconductor detectors.

Chemical influences on the K_β/K_α X-ray intensity ratio of the most energetic lines of an element can be easily observed. In the past there was some confusion, because there is some interference by the kind of excitation of X-rays, which also affects the ratio mentioned. Careful measurements e.g. for proton excitation of X-rays (PIXE) exhibited up to 5 per cent effects.

But recent observations of the secondary X-ray emission of EC nuclides revealed particularly large chemical effects. Up to now neither a clear systematics nor any reasonable explanation of such facts can be found in the literature. We think that we can contribute to this problem by our latest work I will refer to in the following.

If one bears in mind that K_α is a transition in the very inner shell region, the core region, all the effects under consideration will have their origin in influences on K_β. In our opinion there are three main influences on K_β which are related to these problems:

1) The K_β X-ray intensity is caused by 3p − 1s transitions being controlled by the overlap of the 3p and 1s electron states, and this overlap depends on how the screening of 3p by the deeper acting 3d charge is influenced by chemistry. A more or less extended delocalization of 3d charge in the first-row transition element group by effects of "ionic bonding" with its characteristic charge transfer will influence the K_β; the more delocalization takes place, the less is the screening, hence the greater the 3p − 1s overlap and thus the more intense K_β.
This type of influence is characteristic for the first-row transition elements, the 3d elements.
2) Besides this effective charge transfer from the atom to the ligands, in the covalent type of bonding a deformation or so-called polarization of the valence charge distribution around the core takes place. Such a polarization can be interpreted as an admixture of p-type charge to the unpolarized distribution, and as is well known, the dipole selection rules will favour p-type charge in transition to s orbitals ($\Delta l = 1$), here to 1s.
Because of this chemical polarization we have dipole transition possibilities for orbitals which normally cannot give dipole contributions. The higher the degree of polarization, the more intense appears a share of K_β-like radiation by transitions from polarized valence orbitals to 1s.
3) Also under conditions of electron capture a vacancy can be created in the 1s level. But what has changed with respect to other excitation modes? We see mainly two facts: at first, the shielding of 3p by 3d charge is especially high if one has a 3d element, because immediately after the EC has taken place, the daughter atom will possess the electron configuration of the parent atom, that is the atom on the right hand side of the periodic table of elements. And the very fast (10^{-15} s) X-ray emission will happen before this 3d excess charge could relax, and secondly the EC process is conneceted not only with the secondary X-ray emission, but also with a neutrino emitted by the nucleus prior to the X-ray. This neutrino emission causes a recoiling movement of the nucleus with an energy of some eV. Of course, this is a small amount of energy, but within the time interval from γ to X-ray emission, the nucleus has moved about 10^{-9} cm. If the whole electron cloud moved together with its nucleus, nothing special would happen, but will the whole cloud move? We say no! At least such charge fractions which are strongly coupled to their neighbourhood (ligands, next neighbours in lattices ...) cannot move freely, and thus we will get a dislocation of the centres of positive (nucleus) and negative (bound to the ligands) charge, that means we will have an additional type of polarization for reasons of decay kinetics. We will refer to this as "kinetic polarization" in the EC process (see Fig. 2). A somewhat more realistic idea of the geometrical dimensions of this effect is presented in Fig. 3, in which the relative dimensions correspond to the ^{51}Cr situation.

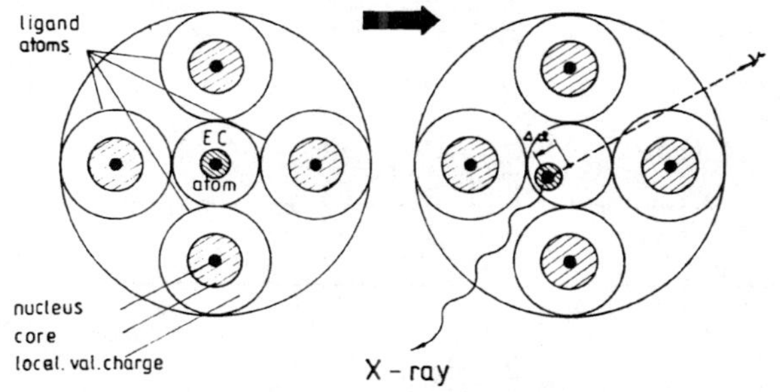

Fig. 2. Kinetic polarization of the central atom by neutrino emission prior to X-ray emission: undistorted molecule (left hand side) and molecule after EC (right hand side)

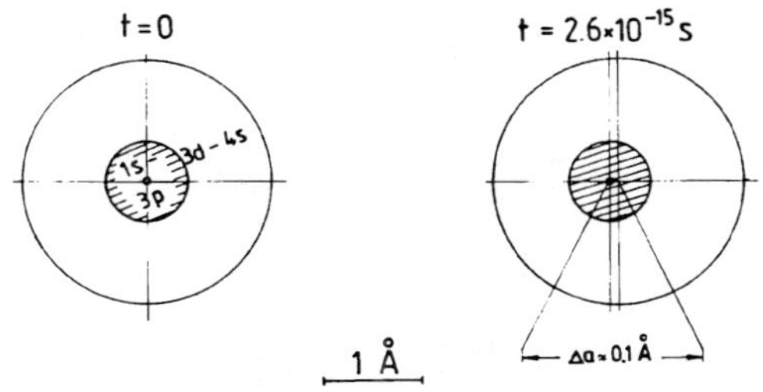

Fig. 3. Kinetic polarization after ^{51}Cr EC

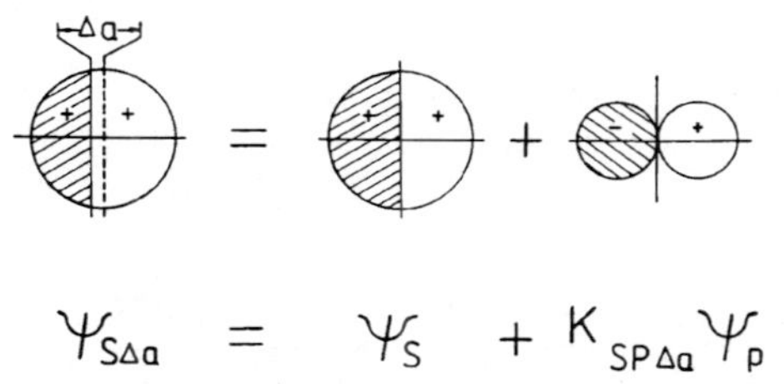

Fig. 4. Development of decentered wave functions
Korrektur: Stat ψ_p lies ψ_P

Whereas the first two arguments (screening and polarization) will find their expression also in ordinary X-ray analytics, this last or third part of the argumentation is specific for using EC atoms as probes for the ligand situation. In the next part of this paper we will deal only with this one.

Some remarks should be made about the respective physics so that suitable use can be made of the phenomenon in analytics.

To start with, one should mathematically represent an electron orbital being decentered with respect to its nucleus as a sum of different but centered orbital functions; e.g. an s-type orbital which is decentered can be described by a centered one plus a p-type function (Fig. 4). In the given connection only additional p-type functions are of interest also for other valence orbital types than s ones in looking at the dipole rules.

An analytical evaluation of the coefficients K has been done (Fig. 5) using hydrogenic wave functions in connection with effective SCF-borne central charges of *Clementi* and *Raimondi*.

The coefficients depend on the orientation of the coordinate system given by the atom-ligand coordination. The c_x, c_y and c_z mean the normalized components of the orbital-nucleus displacement which is expressed as a parallel

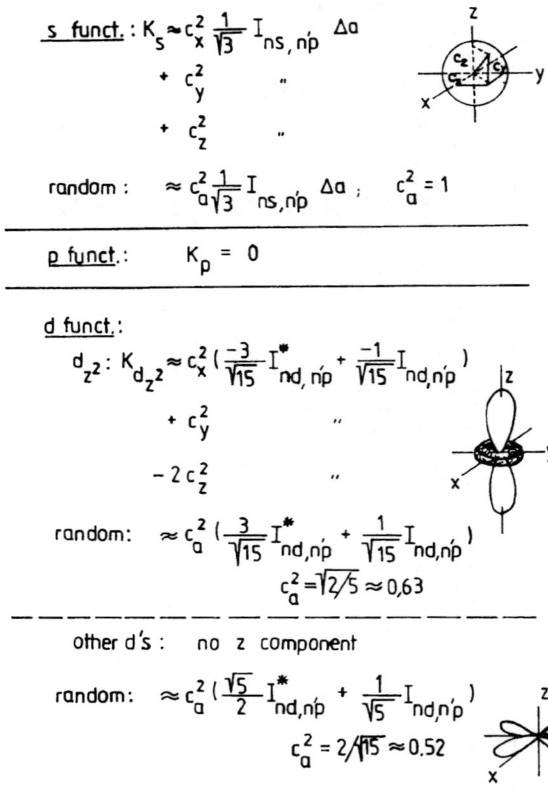

Fig. 5. Coefficients for p-type admixtures in decentered *s*, *p* and *d*-type functions

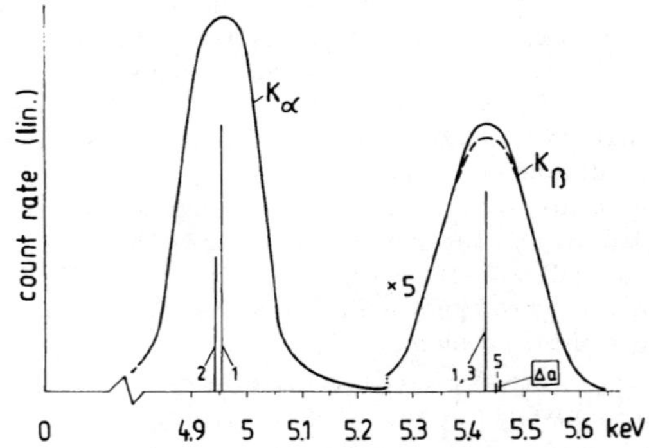

Fig. 7. ⁵¹V KX-rays measured by Si(Li) detector

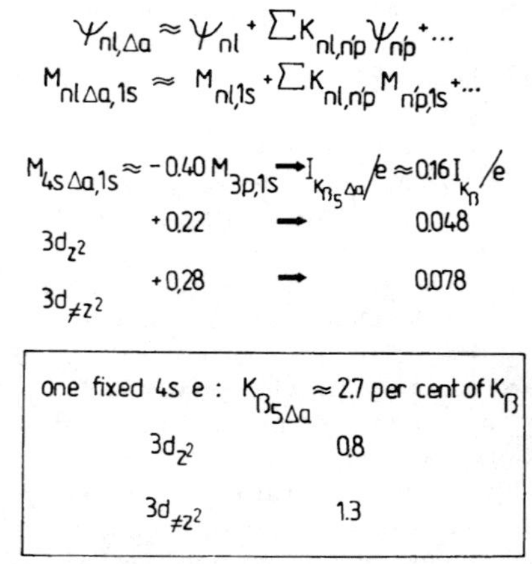

Fig. 8. Dipole transition matrix elements for decentered 4s and 3d orbitals

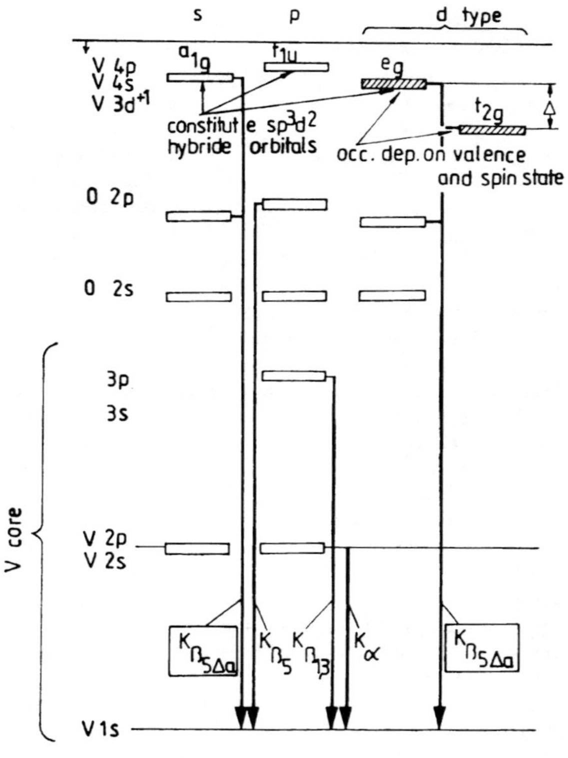

Fig. 6. ⁵¹V KX-rays, oxygen octahedrally coordinated

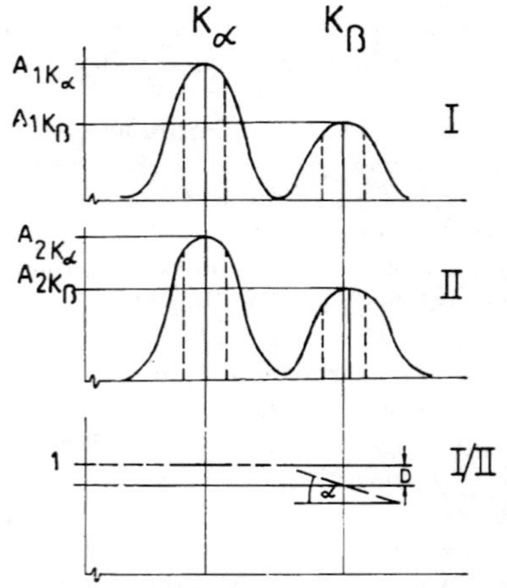

Fig. 9. Spectrogram evaluation by channelwise division

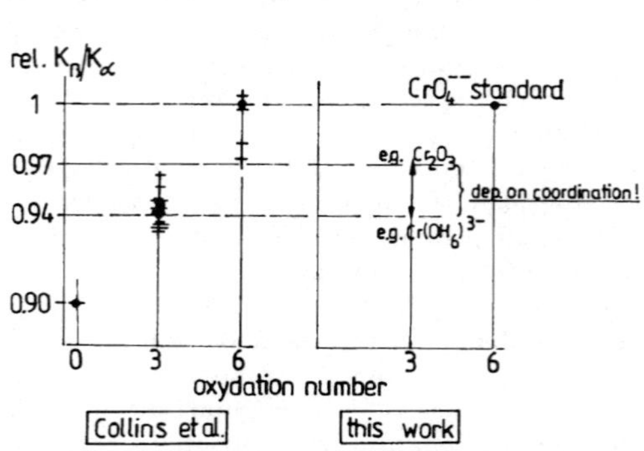

Fig. 10. Kα/Kβ in ⁵¹Cr decay vs. oxidation number

displacement. Of course, the situation depends on the orientation of displacement for the single case. That means a single decay of any EC atom in the sample. But what can be observed is a random case for which statistical mean values of c_x^2, c_y^2 and c_z^2, weighted by the numerical factors (components of the coefficients K), must be applied.

Finally we find kinetic contributions to the K_β intensity, marked by $K_{\beta 5, \Delta a}$ in Fig. 6 from the valence band to 1s, which enhance the total K_β intensity.

What does it look like in the energy dispersive X-ray spectrogram (Fig. 7)? The unresolved K_β peak is somewhat stronger and a little bit displaced into the higher energetic direction, because $K_{\beta 5, \Delta a}$ is added on the higher energetic side ($\Delta E \approx 20 \ldots 30$ eV).

What about the amount of the $K_{\beta 5, \Delta a}$ share?

The last steps in the calculation are as follows (Fig. 8): The displaced wave function $\psi_{nl, \Delta a}$ is written as an expansion of centered wave functions with the K coefficients. The K values do also act as coefficients of the dipole matrix coefficients $M_{nl \Delta a, 1s}$. In this manner the dipole transition matrix elements can be easily written in terms of the matrix elements $M_{3p,1s}$, being responsible for the K_β transitions as a whole.

Regarding the 3p occupation by six electrons one comes to the conclusion that, per one fixed electron of 4s, $3d_{z2}$ and the other four types of 3d orbitals, an increase of the total K_β of 2·7, 0·8 and 1·3 per cent, respectively, should be observed. If somebody were of the opinion that such alteration amounts could not be easily detected, we might refer to our special method for measuring relative K_β/K_α ratios (and small line shifts, too) (see Fig. 9), which is based on the calculation of peak quotients, channel by channel, and application of a least square fit for the quotient values versus channel numbers across channel intervals being nearly equal or somewhat smaller than the peak half widths. From the two parameters of the fitting straight lines the peak amplitude ratio as well as a small relative peak shift ($\ll$ half width) may be drawn.

The transition elements, especially of the 3d row, tend to form complexes, preferably of octahedral symmetry. In terms of the Valence Bond Theory (*L. Pauling*) this means a hybridization in formal combination of two d-, one s- and three p-type atomic orbitals. These hybrid orbitals are directed along the axis' of our coordinate system. Three (namely s, d_{z^2} and $d_{x^2-y^2}$) of the six atomic orbitals involved are sensitive to the mechanism of kinetic polarization in EC events. That is, totally we have a $2{\cdot}7 + 0{\cdot}8 + 1{\cdot}3 = 4{\cdot}8 \approx 5$ per cent effect of increase in K_β under such conditions. But we must bear in mind that this would be a limiting case in so far as a total fixation is assumed of the orbitals involved. In the real bonding situation this condition will not be fully met.

The overall result is that all fractions of charge which are somehow involved in bonding, including those being limited in their ability to follow the nuclear motion, will give rise to contributions of comparable shares enhancing the K_β intensity. From this it becomes clear that taking into consideration also the "normal" chemical effects by delocalization of 3d charge and/or chemical polarization, the chemical effects for EC nuclides are essentially increased. *Collins* et al. [1] found out that a close correlation between K_β/K_α and the oxidation number can be observed in compounds of ^{51}Cr (Fig. 10, left hand side). We are not quite sure if their samples, claimed as "thin", are really thin enough to justify their total neglect of any absorption corrections. But up to now they could not give any reasonable interpretation of this phenomenon, neither of the kind of influence nor of that relatively close correlation.

We known from our theoretical considerations and from our experiments (Fig. 10, right side) that the special coordination of the respective atom must be carefully regarded to predict the K_β/K_α X-ray intensity ratio.

I hope that we have been able to contribute new insights into the interesting feature of EC nuclide decay in bound states which perhaps could lead to some new-type radiotracer applications.

Acknowledgements

Some coworkers of mine are deeply involved in this work. First of all I would like to mention my dear colleague and comrade Dr. *Michael Nagel*, who found an early and unexpected death only a few weeks ago, causing a breach in our work which will be difficult to overcome. But also the work of Dr. *Peter Dostal*, Dr. *Eberhard Hartmann* and *Anette Rigo* should be acknowledged.

From the ZfK Rossendorf Professor *Münze*, Dr. *Johannsen* and Dr. *Mocker* had an essential share in the Tc work.

Received January 19, 1983

References

[1] *K. E. Collins, C. Heitz, J. Cailleret*, J. Chem. Research (M) **1980**, 3401 and J. Chem. Research (S) **1980**, 265

Isotopenpraxis 19 (1983) 9, S. 298 – 301

Measurement of Residence Time Distribution with Radiotracers Using Periodic Pseudo-Random Binary Signal Sequences*)

M. Richter

(Zentralinstitut für Isotopen- und Strahlenforschung der AdW der DDR, Leipzig)[1])

A review about the properties of several types of tracerinput signals is given and in particular the stochastic signals are discussed. An estimation of the necessary measuring time shows, that true stochastic signals are unimportant for tracer experiments. The use of periodic pseudo-random binary signals (PRBS) permits a strong reduction of the measuring time. Therefore their practical application for tracer experiments is possible. The PRBS-method has special advantages in comparison with impulse methods, if fluctuations and disturbances exist within the investigated systems. Correlations for the estimation of parameters, needed for the realization of PRBS-measurements, are given and several measuring arrangements are discussed.

The application of the PRBS-method for measurements of residence time distributions (RDC) of a mixer-settler extractor is described.

Keywords

extraction apparatuses; mathematical models; technetium 99; tracer techniques

The shape of the input signal as function of time influences strongly the contents of information of the output signal in measurements of residence time distributions (RDC) and in other dynamic methods for mathematical modelling and optimization, in which mass flow is studied with radiotracers.

Tab. 1 gives a review about several signal forms having a practical meaning. An analysis of the information contents can be done by calculation of the power spectrums, which are obtained by Fourier transformation. The power spectrums of signals are given in table 1, too. The relation between the power spectrums of the output signal $S_A(\omega)$ is given by Eq. 1:

$$S_A(\omega) = \Phi(j\omega) \times S_E(\omega) \quad (1)$$

$\Phi(j\omega)$ is the frequency response containing the information about the system. Power spectrums and frequency response for two cases are plotted in Fig. 1. In the first case the critical frequency of the input signal $\omega_g(S_E)$ is higher than the critical frequency of the system $\omega_g(\Phi)$.

The output signal contains the complete information about Φ. In the other case the critical frequence of the input signal is lower and the output signal contains no information about the system in the range between $\omega_g(S_E)$ and $\omega_g(\Phi)$.

Tab. 1 shows, that only δ-impulse and the "white noise" signal have an infinite critical frequency. Only by means of these signals a high information is obtained in the experiments. Especially step input or harmonic inputs are unfavourable signal forms.

Whereas impulse inputs are used in most cases, little information is found in the literature about the use of stochastic signals. An analysis of stochastic signals is possible by means of the correlation theory. With a stochastic input signal $x(t)$ and the output signal $y(t)$ the autocorrela-

*) Presented at the 2nd Working Meeting "Radioisotope Application and Radiation Processing in Industry", 28th September to 1st October 1982, Leipzig

[1]) Address: DDR-7050 Leipzig, Permoserstr. 15

Tab. 1. Characteristics of different input signals

signal/typ		power spectrum		critical frequency $\frac{S(\omega_g)}{S(0)} = \frac{1}{100}$
δ - impulse	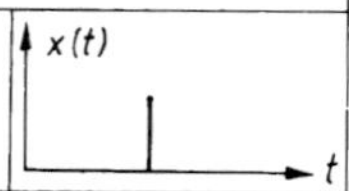	$S_E(\omega) = const.$	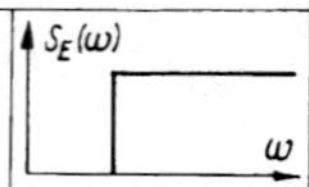	∞
unit step	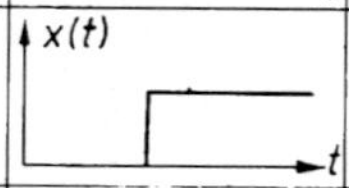	$S_E(\omega) = \frac{1}{\omega}$	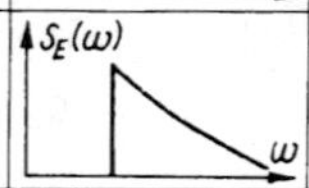	$\omega_g = 100$
harmonic signal (frequency ω_0)	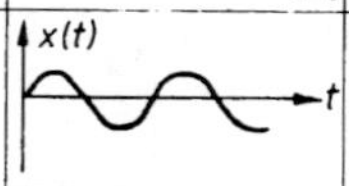	$S_E(\omega) = \delta(\omega - \omega_0)$	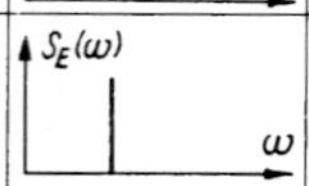	–
white noise	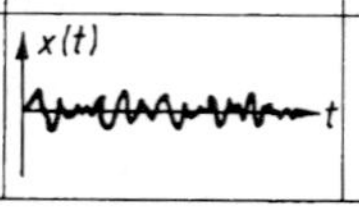	$S_E(\omega) = const.$	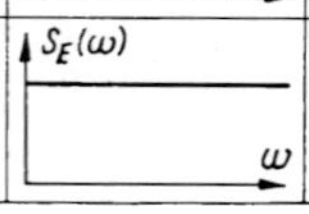	∞

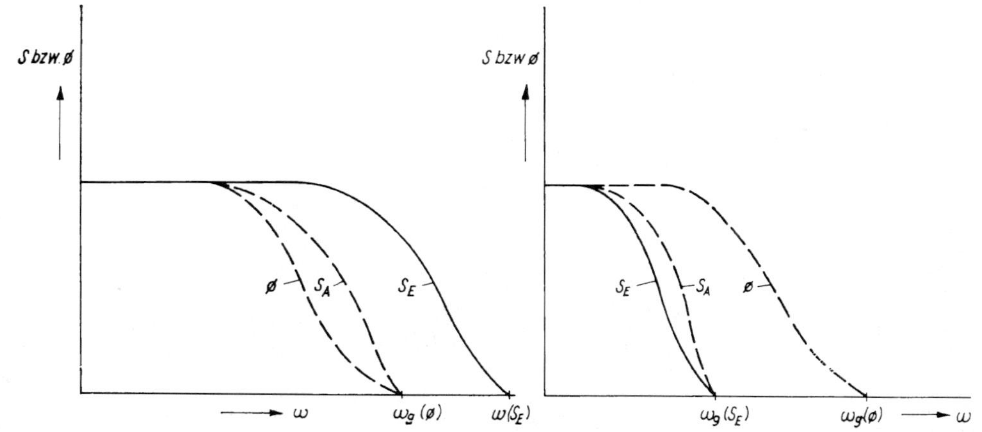

Fig. 1. Power spectrums of the input and output signal (S_E und S_A) and frequency response (Φ) of the system

tion function φ_{xx} and the crosscorrelation function φ_{xy} are defined as:

$$\varphi_{xx}(\tau) = \lim_{T\to\infty} \frac{1}{2T} \int_{-T}^{+T} x(t)\, x(t+\tau)\, dt \tag{2}$$

$$\varphi_{xy}(\tau) = \lim \frac{1}{2T} \int_{-T}^{+T} x(t)\, y(t+\tau)\, dt \tag{3}$$

The input and output signals of a linear system, characterized by the weight function $g(\theta)$, are connected through the following integral

$$y(t) = \int_{-\infty}^{+\infty} x(t-\theta)\, g(\theta)\, d\theta \tag{4}$$

The relation between the auto- and crosscorrelation function is [1]:

$$\varphi_{xy}(\tau) = \int_{-\infty}^{+\infty} \varphi_{xx}(\tau-\theta)\, g(\theta)\, d\theta \tag{5}$$

On base of the similarity of Eq. 4 and 5 the same informations are obtained by using the correlation functions as by using the input and output signals directly.

Eq. 5 is the basic function for the analysis of experiments with stochastic inputs.

The integrals for the calculation of the correlation functions can be practically solved only within a finite interval:

$$\tilde{\varphi}_{xx}(\tau) = \frac{1}{2T} \int_{-T}^{+T} x(t)\, x(t+\tau)\, d\tau \tag{6}$$

For digital computation we use Eq. 7:

$$\tilde{\varphi}_{xx}(\mu) = \frac{1}{N-\mu} \sum_{\nu=1}^{N-\mu} x_\nu x_{\nu+\mu}\,; \qquad \mu = 1, 2, \ldots k \tag{7}$$

The notation for Eq. 7 is shown in Fig. 2.

The total measuring time T, necessary for a relative error of the correlation function lower than $\Delta\varphi$, is given by Eq. 8 [2]:

$$T \geq \frac{1}{(\Delta\varphi)^2} \left(\frac{1}{2\lambda} + \tau_m \left(1 + \frac{\tau}{\tau_m} + e^{\tau/\tau_m} \right) \right) \tag{8}$$

Thereby is λ the mean frequency of the stochastic input signal, which is realized as a sequence of tracer impulses. τ_m is the mean residence time. The number of measured values of a digital computation must be higher than 800000 to get an error lower than 10%. This expense is to high for active experiments.

By use of Pseudo-Random-Binary-Signals, i.e. stochastic impulse sequences repeating after a determined period it is possible however to calculate exactly the correlation function by integration over one period, if no additional noise signals exist within the system. But also with additional noise it is to obtain sufficient accuracy with a much shorter measuring time.

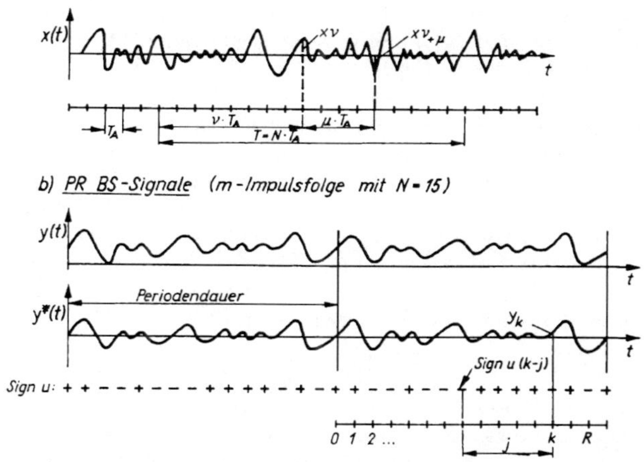

Fig. 2. Relations in the numerical calculation of the correlation function

The relative error of the correlation function ($\Delta\varphi$) for an m-impulse sequence with 127 impulses per period is plotted in Fig. 3 as function of the total number of impulses and the signal-noise-ratio (S_R) [3]. For $S_R \geqq 1$ the error is lower than 10% already after one period.

The m-impulse sequence can be generated by utilization of a back-coupled shifting register. The number of impulses per period N is given by Eq. 9:

$$N = 2^n - 1 \quad \text{(e.g. } N = 15, 31, 63, 127) \tag{9}$$

For much disturbed output signals a low value of N is favourable (e.g. $N = 15$), for little disturbed signals one uses a high number of N (e.g. $N = 127$). Values higher than 127 have no practical meaning, because they require a large tracer activity to obtain a signal-noise ratio larger 1. The length of the period T must be larger than the decay

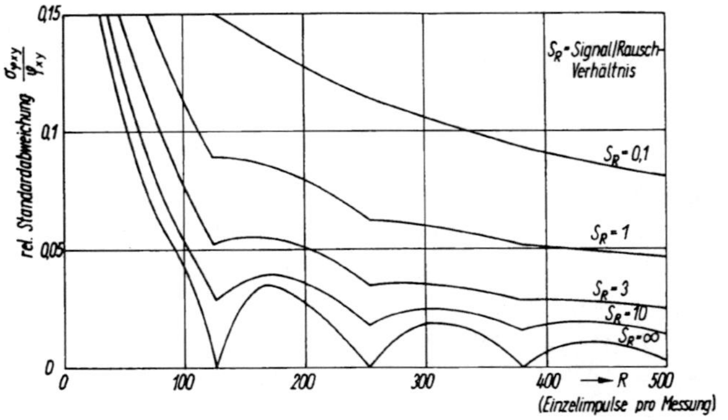

Fig. 3. Influence of the integration period and the signal-noise ratio on the rel. variance of the cross-correlation function for a m-impulse sequence with $N = 127$

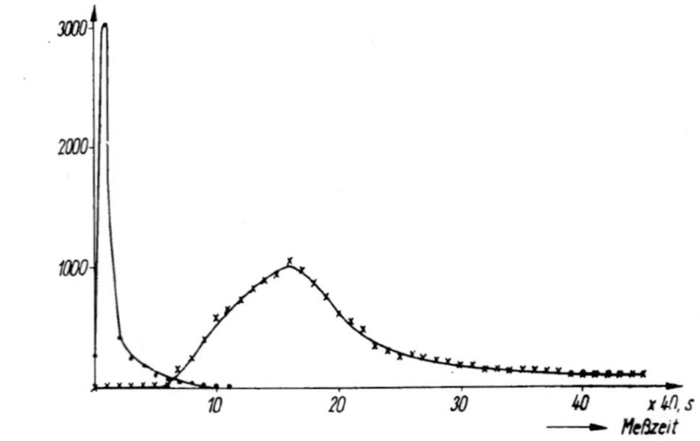

Fig. 4. Residence time measurement with the impulse method
mixer-settler-extractor: 4 stages
1. measuring point: at 1st stage —·—; 2. measuring point: at 4th stage —×—
flow rate: aqueous phase: 1.52 lh^{-1}; organic phase: 0,52 lh^{-1}
tracer: ^{99m}Tc (the same activity as for the PRBS-experiment

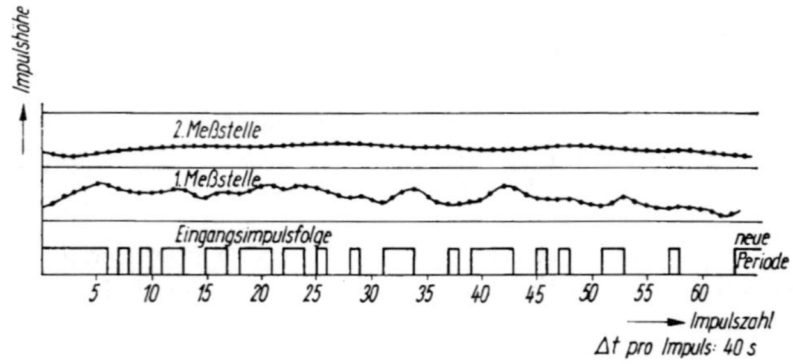

Fig. 5. Measured curves for the RDC-measurement with PRBS-input (m-impulse sequence $N = 63$)
mixer-settler-extractor: 4 stages
1. measuring point: at 1st stage; 2. measuring point: at 4th stage
flow rate: aqueous phase: 1,97 lh^{-1}; organic phase: 0,77 lh^{-1}
tracer: ^{99m}Tc

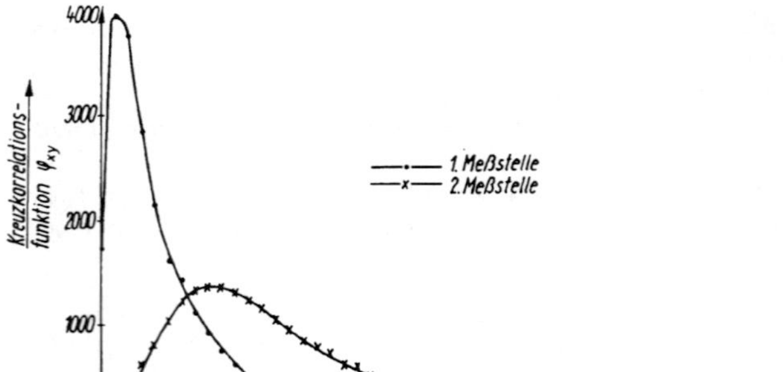

Fig. 6. Crosscorrelation functions of the RDC-measurement with PRBS-method

time of the system. In general this is realized for

$$T \approx 10\tau_m \tag{10}$$

The length of a single impulse is given by

$$\Delta t = T/N \tag{11}$$

The total tracer activity A, necessary to obtain a signal-noise ratio larger 1, is [4]:

$$A \geq \frac{10 \times N}{\tau_m \times \eta} \times P \tag{12}$$

η characterizes the effiency of the measuring arrangement. Equations for the calculation of η for arrangements with outside mounted detectors were derived earlier [5]. P is the number of periods.

Eq. 9–12 permit an estimation of all needed dates for experiments with PRBS-input. For practical application of PRBS-methods it is useful to work with two measuring points recording the input-(y_1) and the output signal (y_2).

For both measuring points the cross-correlation functions are calculated with the impulse sequence function (x) used for the control of tracer supply.

For the Laplace transforms of the two crosscorrelation functions ($\overline{\varphi_{xy}}$) it may be shown [4], that

$$\overline{\varphi_{xy_2}}(S) = \Phi(S)\,\overline{\varphi_{xy_1}}(S) \tag{13}$$

$\Phi(S)$ is the transfer function of the system.

Eq. 13 shows, that after calculation of the correlation functions the same methods can be used for the further analysis as in case of impulse inputs (e.g. method of weighted moments [6]). For stochastic input signals independent from the tracer injection point the same variants of selection of the measuring section are available as for impulse inputs. The autocorrelation function has not to be calculated.

Practically an input sequence in tracer experiments is realized by means of a valve, which interrupts the flow from a reservoir with constant tracer concentration. The input concentration fluctuates between O and a maximum value depending on the concentration in the reservoir and the flow rate. For the calculation of the crosscorrelation function it is necessary to transform the impulse function, so that the time mean-value of the input impulse sequence is zero. This can be done by the folowing transformation:

$$y^*_{ik} = y_{ik} - A_i\,; \qquad A_i = \frac{1}{N}\sum_{k=1}^{N} y_{ik} \tag{14}$$

If the impulse sequences is marked by sign + and −, where +marked the intervals in which the valve is open and −marked the intervals, in which the valve is closed, the correlation function can be calculated by Eq. 15:

$$\varphi_{xy_i}{}^*(j \times \Delta t) = \left(\frac{A_i}{R+1}\sum_{k=0}^{R} \text{sign}\,[u(k-j)] \times y^*_{ik}\right) + \frac{A_i^2}{N} \tag{15}$$

The notation is shown in Fig. 2. The term A_1^2/N must be added, because the autocorrelation function of the m-impulse sequence has the limit $-A_1^2/N$ for $\tau \to \infty$. In experiments the tracer dosing valve is controled correspondening the given impulse sequence, whose parameters were calculated by Eg. 9—11. The best control device is a microcomputer. But it is also possible to use a back-coupled shifting register with variable digits and a variable cycle time. The measured values are recorded digitally with a time interval Δt, estimated by Eq. 11. From this values the crosscorrelation functions are calculated by means of Eq. 14 and 15.

If the tracer concentration calculated by Eq. 12 is used, the measurement is sufficient about one period. But it is to note, that the system needs one period until it obtains the stationary state. Therefore the input signal has to introduce for two periods at least. At the beginning of the second period the record can be started.

The PRBS-method was used for residence time studies in the aqueous phase of a laboratory scale mixer-settler extractor with 4 stages. The tracer was ^{99m}Tc .The injection point was fitted at the feed inlet of the first stage. The measuring points were at the first and the forth stage. In the first experiment we used an impuls input signal. Fig. 4 shows the curves from the measuring points. It is seen, that the output signal has a very long tailing. That refers to an irreversible adsorption of the tracer at the walls and baffles producing an additional disturbance signal. This error could be eliminated by a stochastic input. Therefore in a second experiment a m-impulse sequence was used with 63 impulses per period. The total tracer activity was the same as in the first experiment. The injection took place for two periods. Fig. 5 shows the impulse sequence and the measured curves of the two measuring points. The correlation functions, calculated by Eq. 16, are plotted in Fig. 6. The first curve is a little broader than the impulse curve, because the theoretical autocorrelation function of any PRBS-sequence has a finite breadth. The time delay of the output signal is smaller than the delay of the impuls input method and the tailing is small. This demonstrates that by using of the PRBS-method the adsorption effect pretending a wrong RDC is eliminated. An analysis of the curves plotted in Fig. 4 gives an error of about 100% for the hydrodynamic model parameters in comparison with values calculated from the curves in Fig. 6. The PRBS-method has the advantage, if the output signal becomes strongly distorted by internal fluctuations, if non-linear systems are to describe by linear models or if dynamic process models are to develop. The tracer injection is more complicated, but the total activity is the same as for impuls methods. The accuracy of the PRBS-method is obtained in many cases also by repeated measurements with the impulse method, but the measuring time is much smaller by use of the PRBS-method.

The PRBS-method is an important supplement for the dynamic tracer methods. It will be to advantage if a programmable generator of the wanted impuls sequences would be developed.

Received January 5, 1983

References

[1] *W. W. Solodnikow*, Grundlagen automatischer Regelsysteme Bd. II: Analyse und Synthese linearer Systeme. VEB Verlag Technik, Berlin 1971

[2] *M. Peschel*, MSR (1961) 11, 468

[3] *H. Unbehauen, W. Funk*, Regelungstechnik 22 (1974) 269

[4] *M. Richter*, Dissertation B, AdW der DDR, 1982

[5] *M. Richter*, Isotopenpraxis 12 (1976) 151

[6] *M. Richter*, Chem. Technik 25 (1973) 327

Isotopenpraxis 19 (1983) 9, S. 301—305

Prediction of Mixing Length for Transport of Effluent in Rivers*)

St. Szpilowski

(Institute of Nuclear Research, Warszawa/Poland)[1])

The studies on evaluation of transverse mixing length of effluents discharged into natural streams have been reported. Proposed method bases on measured values of dispersion coefficient. Field investigations were carried out in Vistula river at Warszawa area using radiotracer method. The procedure worked out can be used for prediction of mixing length for sewage outfall under design.

Keywords

radiotracer method; rhodamine B; rivers; wastes dispersion

1. Introduction

Municipal and industrial wastes when disposed to rivers and lakes produce danger for life in surface waters. Continuously increased water pollution level reduces water resources. Pollution of rivers, the main water resources for industry and cities, remains especially difficult problem. One of ways to solve it is seeking methods of waste water discharge providing fast progress of dilution of effluents in ambient water. The most important mechanism of dispersion and dilution of effluents in natural streams is turbulent mixing. The useful tool for investigation of mixing patterns is radioactive tracer method. Its main advantage is that tracer detection remains unaffected by such factors as changes of chemical composition of labelled medium and presence of suspensions. The use of radiotracers in studies of effluent transport in natural streams makes possible to determine factors characterising mixing rate, providing a base for solving problems of optimal location of sewage outfalls.

*) Presented at the 2nd Working Meeting "Radioisotope Application and Radiation Processing in Industry", 28th September to 1st October 1982, Leipzig

[1]) Dept. XVI, ul. Dorodna 16, 03-195 Warszawa/Poland

An important phase of effluent transport in open channel flow is dispersion in the channel reach between the outfall and the point of mixing completing in cross-section of the stream. In this phase the rate of homogenisation process is described by mixing degree. The value of this parameter depends upon intensity of mixing and location of outfall in cross-section.

The purpose of the study being reported is prediction of mixing degree on the base of measured values describing turbulent mixing intencity. The radiotracer method has been used as a tool for measurement procedure of mixing parameters.

2. Theoretical background

The general form of transport equation assuming turbulent flow describes mass balance of substance being transported [1]

$$\frac{\partial c}{\partial t} = \nabla(D \nabla c) - Vc \tag{1}$$

where c is concentration; t is time; D is dispersion coefficient; V is velocity; ∇ is *Laplace* operator.

In equation (1) the dispersion coefficient represents turbulent mixing intensity. The turbulent mixing in open channel flow takes place in longitudinal, transverse and vertical directions.

Because the depth of natural stream is small in comparison to its width and distance necessary to complete transverse mixing, homogenisation process along vertical direction occurs in the period several times shorter than required to complete transverse mixing. The factor controlling mixing in a phase of process before homogenisation completing in cross-section is transverse dispersion [2]. Longitudinal dispersion has to be accounted when the dispersion of instantaneously injected tracer is analysed. Therefore the two dimensional form of equation (1) has been used to study transport process.

The solution of this equation in rectangular co-ordinates, for infinitely wide channel is known in form [3].

$$c(x, y, t) = \frac{A}{4\pi H t (D_x D_y)^{1/2}} \exp\left\{-\left[\frac{(x - ut)^2}{4D_x t} + \frac{y}{4D_y t}\right]\right\} \tag{2}$$

where A is the total amount of instantaneuosly injected substance; H is the depth of a stream; D_x, D_y are longitudinal and transverse dispersion coefficients respectively; u is linear flow velocity; x, y are longitudinal and transverse co-ordinates.

Equation (3) is also valid for dispersion in a region closed to injection point (before the cloud of dispersant reach banks of the channel)

Further analysis of a transport process requires taking into account an effect of reflecting boundaries (banks effect) as well as position of injection point on y-axis. Then equation (1) has form as follows:

$$\frac{\partial c}{\partial t} + u\frac{\partial c}{\partial x} = D_x \frac{\partial^2 c}{\partial x^2} + D_y \frac{\partial^2 c}{\partial y^2} + A\delta(x - x^*)\,\delta(y - y^*)\,\delta(t - t^*) \tag{3}$$

$$x \in (-\infty, \infty)\,; \qquad y \in [0, \mathrm{B}] \tag{4}$$

with initial condition:

$$c = 0 \quad \text{for} \quad t = 0 \tag{5}$$

and boundary conditions:

$$c = 0 \quad \text{for} \quad x \to \pm\infty \tag{6}$$

$$\frac{\partial c}{\partial y} = 0 \quad \text{for} \quad y = 0 \tag{7}$$

$$\frac{\partial c}{\partial y} = 0 \quad \text{for} \quad y = B \tag{8}$$

where A is total amount of injected substance; B is channel width; x^*, y^*, t^* — are coordinates of location and time of injection; δ is *Dirac* function. The equation (3) has been solved by means of Finite *Fourier* Transform [4]. The solution is as follows:

$$c(x, y, t) = \frac{2A \exp\left[-(x - x^* - ut)^2/4D_x t\right]}{B(4\pi D_x t)} \times \left[1 + 2\sum_{n=1}^{\infty} \exp\left(-n^2\pi^2 \frac{D_y t}{B^2}\right) \cos\frac{n\pi}{B} y \cos\frac{n\pi}{B} y^*\right] \tag{9}$$

To analyse the effect of discharge point location and transport time (corresponding to the distance from discharge point) on homogenisation intensity it is convenient to introduce the mixing degree [5]:

$$M = 1 - \left[\frac{1}{N}\sum_{i=1}^{N}\left(1 - \frac{c_i}{\bar{c}}\right)^2\right]^{1/2} \tag{10}$$

where $\bar{c}$ is average concentration in stream cross-section; c_i is the concentration at i-position on the y axis. Equation (10) represents the degree of completion of transverse mixing. $M = 1$ means completion of homogenisation in stream cross-section.

Fig. 1 shows the diagram of mixing degree v.s. dimensionless transport time $D_y t/B^2$ for various discharge point position y^*/B.

The value $y/B = 0$ represents the bank of stream, while $y/B = 0{\cdot}5$ — centerline. The values of $\bar{c}$ and c_i for given y^* have been calculated from equation (9). The mixing length corresponds to the time of transport. Its value may be found from formula:

$$L_m = u t_m \tag{11}$$

It can be seen from Fig. 1. that the rate of homogenisation strongly depends upon discharge point position on y/B axis. The most favourable condition in respect of mixing process occur when discharge point is located in the centerline. Moving the discharge point aside results in decreasing homogenisation rate.

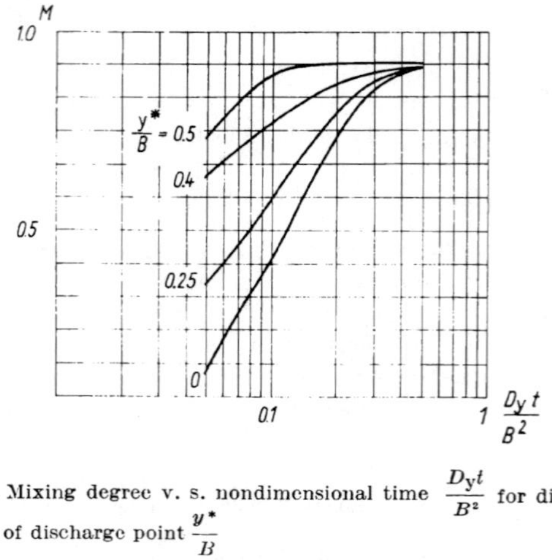

Fig. 1. Mixing degree v. s. nondimensional time $\frac{D_y t}{B^2}$ for different location of discharge point $\frac{y^*}{B}$

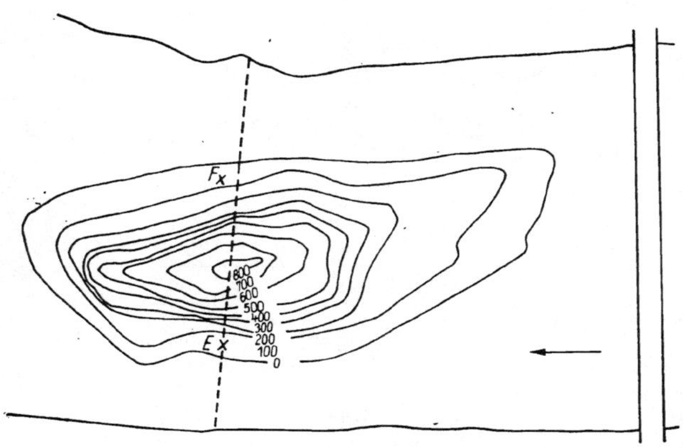

Fig. 2. An example of isoconcentration curves of tracer plume (Run No. 3 — 08. 10. 80)

L = 2425 m; t = 2030 s; u = 1,19 m/s; D_x = 2,03 m²/s; D_y = 0,30 m²/s

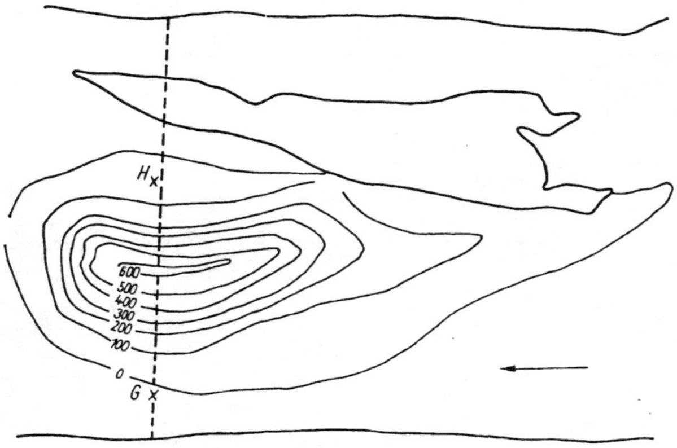

Fig. 3. An example of isoconcentration curves of tracer plume (Run No. 3 — 08. 10. 80)

L = 3175 m; t = 2807 s; u = 1,13 m/s; D_x = 2,15 m²/s; D_y = 0,25 m²/s

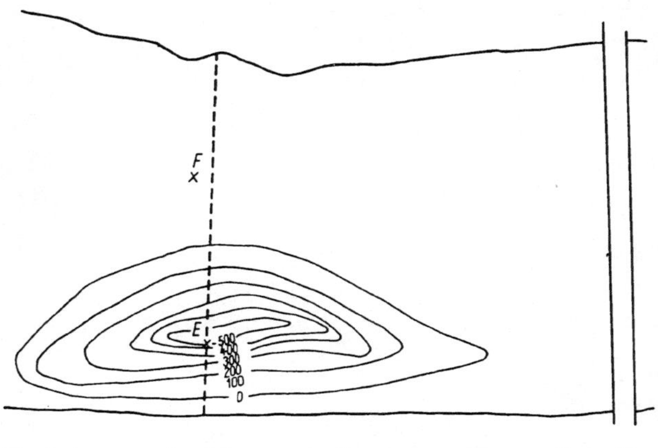

Fig. 4. An example of isoconcentration curves of tracer plume (Run No. 5 — 11. 10. 80)

L = 2335 m; t = 1948 s; u = 1,20 m/s; D_x = 1·79 m²/s; D_y = 0·23 m²/s

3. Field investigation

Described formulas need some confirmation by field experiments. Therefore a series of tracer investigations including measurements of dispersion coefficients and mixing degree values were carried out in Vistula River at Warszawa area.

The dispersion coefficients measurement based on instantaneous injection of tracer at a given point and record-

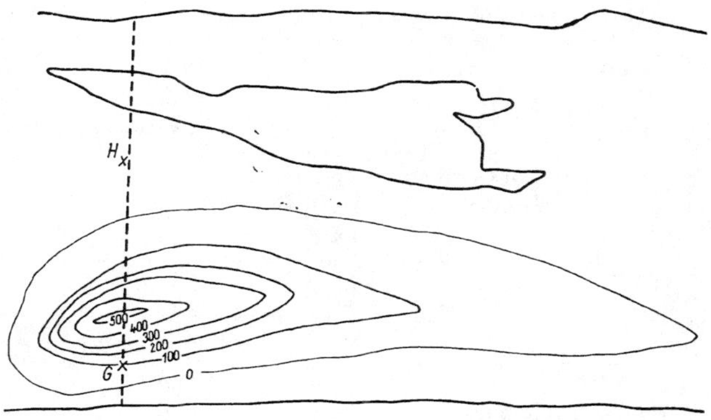

Fig. 5. An example of isoconcentration curves of tracer plume (Run No. 5 — 11. 10. 80)

L = 3085 m; t = 2530 s; u = 1·22 m/s; D_x = 2·48 m²/s; D_y = 0·28 m²/s

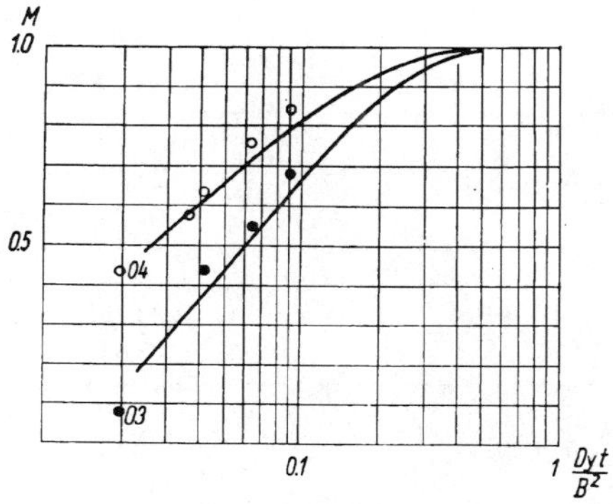

Fig. 6. Measured values of mixing degree and curves $M = f\,[D_y t/B^2]$ calculated for location of discharge points $y^*/B = 0{\cdot}3$ and $y^*B = 0{\cdot}4$

D_y = 0·25 m²/s; B = 347 m; u = 1·11 m/s

ing distribution of tracer concentration along transverse direction. As a tracer ^{82}Br in the form of KBr was used [6, 7]. The curves of distribution of tracer concentration along river width were obtained at measurement profiles located at different distances downstream injection point. The profile was marked by two anchored buoys.

Measuring equipment was installed on the boat with outboard engine. As the tracer plume came between the buoys, transwere cuts were made with the boat. The engine was set at low speed and timed runs were made between the buoys. After the plume had cleared the profile, the cycle was repeated at the next one.

A series of 5 injection was performed using 11·1 GBq (300 mCi) ^{82}Br per one injection. Fig. 2—5 show examples of tracer isoconcentration curves for two injections at various position of injection point on y/B-axis. The values of longitudinal — and transverse dispersion coefficients have been obtained from equation (2) using *Levenspiel — Smith* method [8, 9] modified for open channel flow. The use of this method allows to eliminate error caused by not immediate determination of tracer distribution curves. The results obtained with data of injection point co-ordinates and distances from injection point are shown in Tab. 1.

To check the reliability of equations (9, 10) for predicting mixing degree, the measurements of distribution of continuously released Rhodamine B ware performed. Distribution curves were determined at large (over 20 km) distances from the relase point in the same way as that of the radiotracer tests. As a measuring equipment the Turner

Tab. 1. Data obtained from radiotracer field tests on Vistula river at Warszawa area

Run No.	Injection point coordinates: km of river coarse	Injection point coordinates: distance from the bank	Distance from injection point	Mean flow time	Mean flow velocity	Longitudinal dispersion coefficient	Transverse dispersion coefficient
		y	L	t	u	D_x	D_y
		m	m	s	m/s	m^2/s	m^2/s
			560	479	1.17	2.12	0.24
1	519.62	62	980	860	1.14	1.86	0.27
			1360	1204	1.13	2.07	0.31
			2240	2000	1.12	2.05	0.22
			580	487	1.19	1.74	0.32
			960	800	1.21	1.81	0.26
2	519.80	112	1360	1192	1.14	1.63	0.24
			2320	1982	1.17	1.57	0.28
			510	*)	—	—	—
3	511.86	125	2425	2030	1.19	2.03	0.30
			3175	2807	1.13	2.15	0.25
			420	355	1.18	1.78	0.26
4	511.88	45	2415	2127	1.13	1.99	0.13
			3165	2857	1.11	2.75	0.21
			610	*)	—	—	—
5	511.95	55	2335	1948	1.20	1.79	0.23
			3085	2530	1.22	2.48	0.28

*) Not reliable measurement of buoy position

Tab. 2. Measured values of mixing degree and respective reduced values

Run No.	Release point co-ordinate	Distance from release point	Nondimensional time	Mixing Degree: Measured	Mixing Degree: Predicted
	y/B	L	$D_y t/B^2$	M_m	M_p
		m			
		19250	0.036	0.57	0.578
1	0.4	21920	0.041	0.63	0.605
		33150	0.062	0.76	0.703
		48110	0.090	0.84	0.778
		21250	0.041	0.43	0.3777
2	0.3	33150	0.062	0.54	0.512
		48110	0.090	0.68	0.619

Model 10-005 fluorometer (continuous flow equipment) was used. Two experiments were performed. The results are given in Tab. 2. In calculations of mixing degree the averaged measured values of dispersion coefficients has been used. Fig. 6 shows diagram of mixing degree v.s. dimensionless time $D_y t/B^2$ and respective measured values.

4. Conclusions

Suggested procedure of measurement of dispersion coefficient and prediction of mixing degree disregards effect of irregularities of bottom and banks shape. Under such conditions measured values of dispersion coefficients can be overestimated. Measured values of dye mixing degree (Fig. 6) show small positive deviation from calculated values. It may be assumed that existing irrgularities of river bed have insignificant effect on error of predicting procedure when measured values of dispersion coefficients are used. It seems that main effect of existing irregularities is promoting mixing process, what can be seen from measured values of mixing degree which are higher than respective predicted values.

The results obtained lead to the conclusion that presented method of predicting mixing degree on the base of measured values of dispersion coefficients can give the reliable results for the purposes of choosing optimal location of sewage outfalls.

The author is of the opinion that further work should include investigation of dispersion of effluent in close vicinity of discharge point and effect of design features of outfall on initial dilution.

Received January 26, 1983

References

[1] *V. L. Streeter*, Handbook of Fluid Dynamics, New York – London 1961
[2] *D. M. Ruthven*, Water Research **5** (1971) 343
[3] *A. N. Diachishin*, J. San. Eng. Div., Proc. ASCE **89**, 1 (1963) 29
[4] *I. N. Sneddon*, The Use of Integral Transforms, McGraw-Hill Book Comp. Ltd. (1972)
[5] *P. R. Ward*, J. Hydr. Div., Proc. ASCE **99** (1973) 1069
[6] *K. R. Mitchell, G. E. Brendish*, J. Inst. Sewage Purification **6** (1962) 586
[7] *W. Zabłotny, R. Winnicki*, Nukleonika **16** (1971) 389
[8] *O. Levenspiel, W. K. Smith*, Chem. Eng. Sci. **6** (1957) 227
[9] *St. Szpilowski, G. Strzelczak, R. Winnicki*, Nukleonika **21** (1976) 603

Isotopenpraxis **19** (1983) 9, S. 305–307

Crosslinking and Free Radical Decay Reactions in Irradiated Polyethylene in Presence of Additives*)

E. Jaworska, I. Kaluska, G. Strzelczak

(Institute of Nuclear Research, Warsaw)[1])

The blends of low density polyethylene and additives of flame retardants type were irradiated in a ^{60}Co gamma source and with 13 MeV electrons from linac, at temperature of environment about 20 °C. Esr measurements were performed just after irradiation and after storage at room temperature.

It was stated that dominated radical present in the sample depends on the additives, dose and storage time. With increasing dose the overall concentration of radicals in samples irradiated with electrons-decrease. The correlation between ability to allyl radicals formation and efficiency of crosslinking is explained by differences of radicals mobility in the presence and absence of additives.

Keywords

cross-linking; dose rates; electron beams; electron spin resonance, gamma radiation; organic bromine compounds; polyethylenes, radicals

1. Introduction

Radiation induced reactions in polyethylene have been a subject for a large number of works. Special attention had been paid to the formation and decay mechanism of reactive sites that are formed. Most of them concern radical reactions and have used esr as experimental technique [1–5]. To understand better the reactions in that semicrystalline polymer, polyethylene with a definite structure, i.e. with known molecular weight distribution, number of end vinyl groups, crystallinity and crystal size had been used [6, 7]. In some cases the model systems of n-paraffins had been studied [8] or reactions in isolated polymer chains [9]. For that reason investigation of polyethylene blends is more complicated. The temperature and time period of processing effects significantly the scission of polymer chains, appearance of some constituents like oxidation products and unsaturation [10]. The presence of additives in the blends changes process of polyethylene crystallisation and structure of interface zones. The crystal size and concentration of excluded vinyl groups $>C{=}CH_2$ might be expected to be different from that of pure polyethylene. All that factors, as well as the temperature influence any radical reactions and efficiency of end products [6].

Our previous results [11] indicate, that gel content of blends irradiated with fast electrons are lower then of those irradiated in gamma source and that both values are lower for pure polyethylene. Suspection, that additives, which are aromatic compounds, deactivate active centers created on polymer molecules and hinder radiation induced crosslinking, has not been confirmed by radical yield measurements. The *G* values of alkyl radicals (calculated from esr spectra at l.n.t.) for blends were higher then for polyethylene. In the presence of additives radicals seems to be localized in the interfacial zone of the polymer.

Because at l.n.t. any radical reactions are suppressed the work has been undertaken in an effort to elucidate radicals reactions in samples irradiated at temperature of environment about 20 °C and then stored at room temperature. In fact the temperature of irradiation was higher and depended on heat capacity of the components. The estimated increase of temperature of sample components irradiated with fast electrons is shown in the Tab. 1. The temperature of irradiation in ^{60}Co source was much lower, because of more effective heat exchange with environment. In the calculations the increase of heat capacity with temperature had not been taken into account.

2. Experimental

The flame retardants used in the study were as follows: tetrabromobisphenol A (TBD), octabromodiphenyl ether (OBDE), and antimony trioxide Sb_2O_3. The blends were prepared from LDPE ($d = 0{\cdot}9209$, F.I. $= 0{\cdot}3$), 3% Sb_2O_3 and 10% of OBDE or TBD mixed at 150–160 °C for 15 min and extruded at 170 °C .The films were cut to appro-

*) Presented at the 2nd Working Meeting "Radioisotope Application and Radiation Processing in Industry", 28th September to 1st October 1982, Leipzig

[1]) Address: Dorodna 16, 03-195 Warsaw, Poland

Tab. 1. Estimated increase of temperature of blend components irradiated with fast electrons

Component	*Cp*,	Increase of temperature, °C		
	J/g	30 kGy	60 kGy	100 kGy
polyethylene	1·88	16	32	53
bromine-compounds	0·67	45	90	150
Sb_2O_3	0·33	90	180	300

priate size (35 mm × 2 mm × 0·5 mm) evacuated and irradiated in ^{60}Co source and with 13 MeV electrons from linac. The dose rate was 4 kGy/h and 1·5 × 10^4 kGy/h, respectively. The esr spectra were recorded with the X-band JES-ME-3X JEOL spectrometer and 100 kHz magnetic field modulation. The microwave power was 2 mW. The calculation of spin concentration was carried out graphically from central peak heights and compared with standard samples. Concentration of radicals had been recounted for pure polyethylene.

3. The results and discussion

The ESR spectra of polyethylene and its blends irradiated with fast electrons are shown on Fig. 1 and Fig. 2. The spectra had been taken several minutes after irradiation — Fig. 1a, 2a and after 1 day storage at room temperature — Fig. 1b, 2b. As seen from Figures the type of dominated radical present in the sample depends on the additives, dose and storage time.

Tab. 2. Luminescence data of bromine compounds [13]

Component	Emission energy, cm^{-1}
tetrabromobisphenol A	25 088
decabromodiphenyl ether	23 226

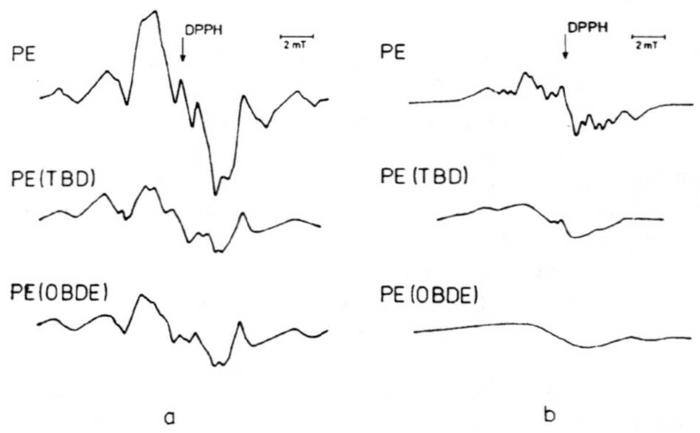

Fig. 1. ESR spectra of samples irradiated with electrons up to 35 kGy (a) and 130 kGy (b)

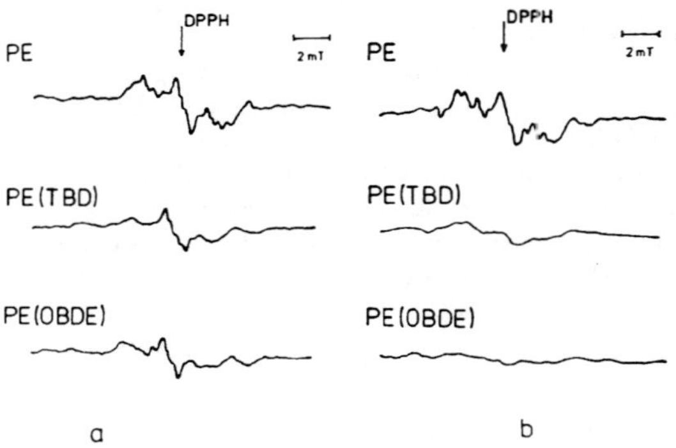

Fig. 2. ESR spectra of samples irradiated with electrons up to 35 kGy (a) and 130 kGy (b) after 1 day storage at room temperature

All spectra of pure polyethylene carried out just after irradiation, with exception for low dose, are typical for allyl radicals.

The spectra of blends for low dose and zero storage time are the mixture of alkyl and allyl radicals, but alkyl radicals are in majority. After storage allyl radicals are going to dominate and the wing peak characteristic for alkyl radicals is no more distinguishable. With increasing dose the overall concentration of radicals decrease, but the shapes of the spectra are going to be changed.

We believe, that formation of allyl radicals in blends process trough the capture of alkyl radicals by vinylen double bonds. Formation of alkyl radicals by energy transfer from excited aromatic molecule to carbon-carbon double bond in polyethylene and elimination of an allylic hydrogen atom to form allylic radicals [12] seems to be, for that types of additives, energetically not possible. From comparison of emission energy of brominated compounds — Tab. 2 and absorption energy of double bonds of polyethylene, about 185 nm = 54054 cm^{-1} one can see, that energy level is too low to excite the double bond of polymer.

That calculations are in line with results of gel content. The formation of allyl radicals on the other pathway as alkyl radicals decay ought, in consequence, to increase gel content of the blends, but it has not been observed [11].

The concentration of radicals in samples irradiated with electrons and in gamma source, then stored for 1 and 5 days at room temperature, are presented in Tab. 3.

It is seen from the Table, that in all samples irradiated with low dose and then stored at room temperature, allyl radical dominate. The concentration of radicals in samples irradiated with gamma rays is much more enhanced. For the dose 70 kGy and higher, the concentration of allyl

Tab. 3. Concentration of radicals*), in relative numbers (±0·05), present in samples after 1 and 5 days storage time

Sample	Dose, storage time							
	e						γ	
	35 kGy		70 kGy		130 kGy		15 kGy	
	1 d	5 d	1 d	5 d	1 d	5 d	1 d	5 d
PE	1·35	1·1	0·38	0·36	0·31	0·30	1·1	0·9
	a	a	a	a	a	a	a	a
PE (TBD)	0·25	0·22	0·15	0·15	<0·1	<0·1	0·48	0·41
	a	a	b	b	b	b	a	a
PE (OBDE)	0·18	0·15	<0·1	<0·1	<0·1	<0·1	0·53	0·42
	a	a					a	a

* dominated radical: (a) allyl, (b) bromine compound

radicals in polyethylene decreases. In blends allyl radicals are not observed. The residual radicals in blend with TBD seems to be possibly the radical of flame retardant, stable at room temperature.

From the comparison of results presented in the Tab. 2 and gel content [11] it is seen, that there is a correlation between ability to allyl radicals formation and efficiency of crosslinking. It can be explained by differences of radicals mobility in the presence and absence of additives and different dependence of G (crosslinking) and G (unsaturation) values on the temperature. Efficiency of allyl radicals formation is influenced by the size of crystalites. In the absence of additives, the order of the system might be expected to be higher then that for blends and formation of allyl radicals is pronounced. In the presence of additives, the radicals are formed preferable in the interface zone, their mobility is high even below α-transition temperature [14]. For that reason polymer radicals are not observed in blends irradiated above 70 kGy. The yield of end products depends on the temperature. Because of higher dose rate (= higher irradiation temperature) decrease of the yield of crosslinking was observed. It means, that formation of double bonds occurs faster then interchain bonds. That conclusion is in line with the other literature data [6, 15]. Confirmation of that hypothesis would be determination of double bonds concentration in blends irradiated at various conditions.

Received January 26, 1983

References

[1] *A. Charlesby, D. Libby, M. G. Ormerod,* Proc. Royal Society, A (1961) 262, 207
[2] *D. C. Waterman, M. Dole,* J. Phys. Chem. 74 (1970) 1913
[3] *J. Auerbach, L. H. Sanders,* Polymer 10 (1969) 579
[4] *L. K. De Vries, R. H. Smith, B. M. Fanconi,* Polymer 21 (1980) 949
[5] *T. Fujimura, N. Hayakawa, I. Kuriyama,* Polymer 19 (1978) 1031
[6] *K. Mandelkern* in "The Radiation Chemistry of Macromolecules" ed. *M. Dole A.P.*, N.Y. 1972, Vol. 1, 287
[7] *G. N. Patel,* J. Polymer Sci. 13 (1975) 351
[8] *W. J. Chappas, J. Silverman,* Radiat. Phys. Chem. 16 (1980) 437
[9] *S. Shimada, Y. Hori, H. Kashiwabara,* Polymer 18 (1977) 25
[10] *M. U. Amin, G. Scott, M. K. Fillekaratne,* Eur. Polymer Sci. 11 (1975) 85
[11] *E. Jaworska, I. Kaluska,* in "Proceedings of 5th Tihany Symposium" 1982
[12] *T. Takeshida, K. Tsuji, T. Seiki,* J. Polymer Sci., A-1, 10 (1972) 2315
[13] *T. E. Dawidson, C. W. Roberts,* J. Appl. Polymer Sci. 25 (1980) 2439
[14] *S. Shimada, Y. Hori, H. Kashiwabara,* Polymer 22 (1981) 1377
[15] *M. Imai,* Radiat. Phys. Chem. 18 (1981) 425

Isotopenpraxis 19 (1983) 9, S. 307 – 309

Substoichiometric Determination of Pd-II-Traces with Dithizone and Problems of Determination*)

H. Wagler

(Academy of Sciences of the GDR, Central Institute of Isotope and Radiation Research, Leipzig)¹)

Pd-traces determination by radioactive isotope dilution analysis based on solvent extraction of metal chelates — Dithizone is a reagent frequently used for the substoichiometric separation of trace amounts of metals. In the case of substoichiometric Pd-extraction there are some problems in the reproducibility of determination, because in solutions of dithizone in organic solvents exist two tautomeric species. At present it is supposed in the literature that only the thiol form of dithizone reacts with the metal ions. Extensive studies of reaction between Pd(II) and dithizone using ^{130}Pd *suggest an extraction mechanism with the thione form. From this fact follow the general problems and the peculiarity of Pd-determination.*

Keywords

accuracy; chelating agents; dithizone; equilibrium; isotope dilution; Palladium; substoichiometry; trace amounts; tracer techniques

1. Introduction

Substoichiometric isotope dilution analysis based on solvent extraction of metal chelates is a simple but very selective, sensitive and reproducible method for determination of trace amounts of metals [1]. Presupposition for this radiochemical method is the existence of a radioisotope of high specific activity of the metal in question and the existence of a suitable organic reagent. There is only a small quantity of suitable organic reagents. Dithizone already has been used for the determination of a large number of metals [2, 3], the Pd-determination was published too [4]. However in solution of dithizone in organic solvents two tautomeric species (thione-thiol-tautomeric) exist. Up to now it is supposed in the literature that the tautomeric equilibrium of dithizone has no importance for the extraction mechanism and that only the thiol form is able of reacting with metal ions [5, 6].

But the problems with the reproducibility of Pd-extraction by means of dithizone suggested that Pd-ions react preferably with the thione form. Extensive studies support this supposition [7].

2. Problem: Application of a tautomeric reagent for substoichiometric determination

Substoichiometric separations are carried out with less amount of reagent than required for complete reaction with the total amount of metal:

$$M^{n+} + n(HA)_{org} \rightleftharpoons (MA_n)_{org} + nH^+$$

$$(M^{n+}) > (HA)$$

$$K = \frac{(MA_n)_{org}(H^+)^n}{(M^{n+})\,(HA)^n_{org}}$$

Assumption: more than 99% of the organic reagent (HA) added are to form the metal chelate (MA_n).

*) Presented at the 2nd Working Meeting "Radioisotope Application and Radiation Processing in Industry", 28th September to 1st October 1982, Leipzig

¹) DDR-7050 Leipzig, Permoserstr. 15

This means for dithizone that the tautomeric equilibrium shifting is necessary, i.e. a shifting to the tautomeric form reacting with the metal in question.

Dithizone (Diphenylthiocarbazone H_2D_z):

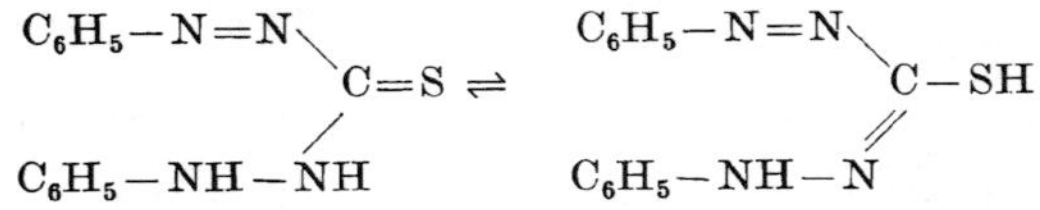

thione form — thiol form

Tautomeric equilibrium shifting is possible by selecting the suitable extraction conditions:

organic solvent for dithizone	tautomeric equilibrium shifting **thione form** less than 100%	**thiol form** to 100%	medium of aqueous phase
chloroform	↑		acid
carbon-tetrachloride			
cyclohexane		↓	alkaline
reaction with	palladium	e.g. copper	

Note: The tautomeric equilibrium shifting to thione form does not go to completion!

3. Experimentals-Development of method

Radioisotope: carrier free ^{103}Pd, $T_{1/2} = 17$ d
^{64}Cu, $T_{1/2} = 12.9$ h)

Apparatus: Scintillationscounter with NaJ(Tl)-crystal, *p*H-meter MV 84 (Precitronik, Dresden), mechanical shaker

Stoichiometry and reproducibility

Substoichiometric determination is based on activity comparison of equal substoichiometrical quantities isolated from a standard (A_{st}) and the sample (A).

The amount of the element to be determined (x) was calculated according to the equation: $x = y\left(\frac{A_{st}}{A} - 1\right)$.

(The same amount of ^{103}Pd (y) was added to the standard and to the sample and it was extracted with same amount of dithizone).

Determination of the region of the constant reaction ratio for estimation of the metal reagent rate for the standard

a) *Pd-extraction* in acid solution
by means of dithizone in chloroform:
metal chelate: PdXHDz
stoichiometry: $Pd^{2+}/H_2Dz = 1:1$
According Fig. 1: For 100% shifting to the thione form thus a complete reaction metal excess is necessary:
0.1 N H_2SO_4: 2-fold
0.1 N HCl : 3-4-fold (Pd-chlorocomplexes in aqueous phase).

b) *Compared to*: *Cu-extraction* in neutral solution
by means of dithizone in cyclohexane:
metal chelate: CuDz
stoichiometry: $Cu^{2+}/H_2Dz = 1:1$
According Fig. 2: By the given reaction conditions dithizone exists already in the thiol form (100%) therefore the region of the constant reaction ratio is attained at about stoichiometric metal reagent ratio.

Cu-II-determination is limited alone by specific activity of the radioisotope.

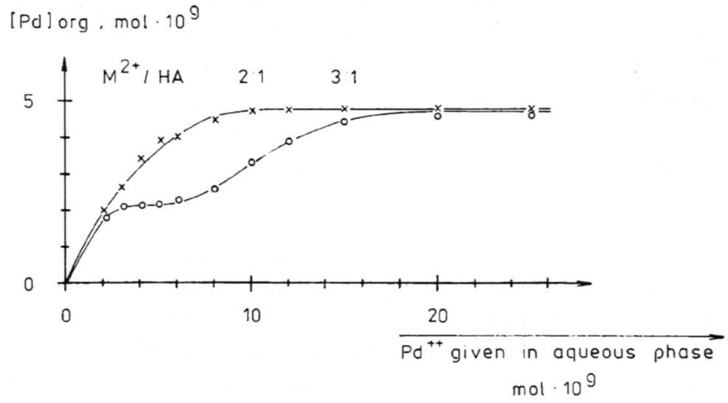

Fig. 1. Pd^{++}-extraktion by means of 5×10^{-9} mol $H_2Dz/CHCl_3$

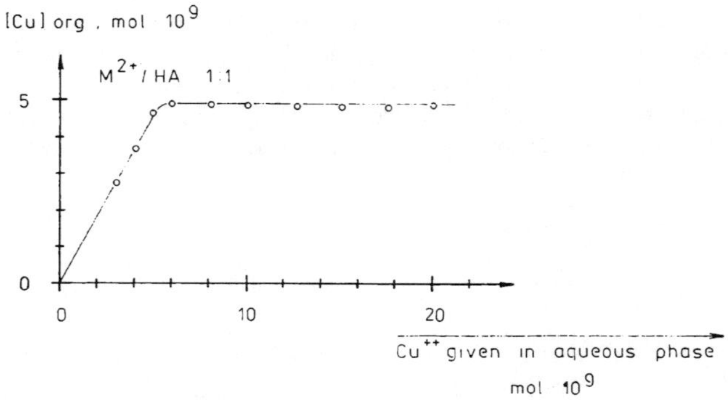

Fig. 2. Cu^{++}-extraction by means of 5×10^{-9} mol H_2Dz/C_6H_{12}

But by the Pd-determination there are some problems because a metal excess is necessary for completion of reaction and also necessary for a lowering of detection limit.

Lowering of detection limit

Call to: $(M^{n+}) > n(HA)_{org}$

A general possibility is the calculation of the corresponding *p*H-values at which a required amount of metal can be extracted:

$$pH > \frac{1}{n}\log\frac{C_{HA}}{n} - \frac{1}{n}\log\left(C_M - \frac{C_{HA}}{n}\frac{V_{org}}{V}\right) - \frac{1}{n}\log K - \log 0.01 C_{HA}$$

the equation is applicable only when the dissociation of the organic reagent in the aqueous phase can be neglected. A lowering of detection limit can be realized by increase the extraction-*p*H-values,

provided, no changes in the reagent by *p*H increase!

The *p*H dependence of the tautomeric equilibrium of dithizone means for the Pd-extraction a decrease of the reaction. A lowering of detection limits realizable by increase of the metal excess only:

trace amounts range palladium	metal excess necessary for the standard aqueous phase: H_2SO_4	HCl
2 to 0,2 μg/ml	2-3-fold	3-4-fold
0,2 to 0,02 μg/ml	3-4-fold	>5-fold
0,02 to 0,002 μg/ml	>5-fold	

Note: Common and favourable metal excess for the standard: 2-fold!

Example for the reproducibility and accuracy of Pd-determination

An example is given in Tab. 1.

Tab. 1. Determination of Pd-traces in the range of about 1 μg to 0.1 μg

^{103}Pd added mol $\times 10^9$ (y)	palladium mol $\times 10^9$ (x)	reagent H_2Dz/ $CHCl_3$ mol $\times 10^9$	activity of org. extract cpm	palladium found mol $\times 10^9$ $x = y \times \left(\frac{A_{st}}{A} - 1\right)$	deviation %
5.0	—	1.5	43200	standard	—
5.0	1.25	1.5	34200	1.31	+4.8
5.0	2.5	1.5	28920	2.46	−1.6
5.0	5.0	1.5	21670	4.98	−0.4
5.0	10.0	1.5	14450	9.95	−0.5
5.0	12.5	1.5	13395	12.43	−0.6

V/V_{org} = 5/1.5; extraction time: 30 min.

Pd-traces in the range of 0.2 μg/ml to 0.02 μg/ml can be determined with good reproducibility and accuracy by means carrier free ^{103}Pd. However the determination of extremly small amounts of metal may often be difficult, because small amounts of metal means still lower reagent concentration. To fulfil the conditions of isolating 99% MA_n or exactly the equal amounts of MA_n from standard and sample, is more difficult because of unstabilities of the reagent at low concentrations. The necessary high metal excess for the Pd-determination means also a very small reagent concentrations.

4. Conclusion

From the necessary high metal excess in the case of Pd-determination in the region of constant reaction ratio it follows an unfavourable rate of given activity in the aqueous phase to measurable activity in the organic phase.

Even thougth a carrier free radioisotope is used it may happen that the error of calculation is very large.

A new approach is needed: An alternative solution is the substoichiometric determination in the region of non-constant reaction ratio by evalution using calibration curves. Then only a small metal excess is necessary.

It is not any longer necessary that the organic reagent react at 99% or at equal parts, only demanded: formation of just one definite metal chelate under given reaction conditions.

Trace amounts of Pd-II below 0.01 μg/ml can be determined with a good precision by the method based on the determination in the region of non-constant reaction ratio by evalution using calibration curves.

5. Application:

Cu-II-determination in semiconductor materials,
Pd-II-determination in catalyst materials.

Received January 19, 1983

References

[1] *J. Ruzicka, J. Stary*, "Substoichiometry in Radiochemical Analysis", Oxford: Pergamon 1968
[2] *K. Kudo, N. Suzuki*, J. Radioanal. Chem. 26 (1975) 327
[3a] *J. Ruzicka, J. Stary*, Talanta 8 (1961) 228
[3b] *J. Ruzicka, J. Stary*, Talanta 9 (1962) 617
[4] *G. B. Briscoe, S. Humphries*, Talanta 18 (1971) 39
[5] *G. Iwantscheff*, „Das Dithizon und seine Anwendung in der Mikro- und Spurenanalyse“, Weinheim: Verlag Chemie 1972
[6] *H. M. N. H. Irving*, "Dithizone", Anal. Sciences Monograph No. 5, Chem. Soc. London 1977
[7] *H. Wagler*, D. thesis, Leipzig 1981

Isotopenpraxis 19 (1983) 9, S. 309–312

Fast Analysis of Element Contents by Inelastic Scattering of Neutrons and the Method of Time Correlated Associated Particles*)

B. Heinrich, G. Musiol, U. Richter

(Dresden University of Technology, Department of Physics)[1])

The element content measurement programme of the Dresden University of Technology, Department of Physics, using the γ-radiation after inelastic scattering of 14 MeV neutrons is described. To suppress the intense background the **T**ime **C**orrelated **A**ssociated **P**article **M**ethod *(TCAPM) is used. First experimental results of carbon, silicon and oxygen content measurements with mixtures of graphite and sand are given. The use of this method for determining the calorific value of brown coal is discussed.*

Keywords

coal; fast neutrons; inelastic scattering; quality control

1. Introduction

Quantitative analysis are introduced in all branches of industry today. This is necessary for selection and control of raw materials, investigation of the best composition of materials and the examination of intermediate and finished products and offal. Chemical analysis have the disadvantage that the analysing time is often in the hour range. For instance the analysis of coal needs a time interval of about 2–3 hours. That means that the results are available after the materials are worked up. Thats why new methods are developed at the Dresden University of Technology, Department of Physics, which allow fast as well as exact element content analysis. The method, described here, is an

*) Presented at the 2nd Working Meeting "Radioisotope Application and Radiation Processing in Industry", 28th Sept. to 1st Oct. 1982, Leipzig

[1]) Anschrift: 8027 Dresden Mommsenstr. 13

element specific one suitable for determination of many elements [1]. As an example we show the application of inelastic scattering of 14 MeV neutrons coupled with the Time Correlated Associated Particle Method (TCAPM) for the determination of carbon, silicon and oxygen in simulated brown coal samples.

2. On-stream analysis in brown coal industry

With about 300 mill. t/year the GDR is one of the most important brown coal mining states of the world. Therefore the importance of methods for coal parameter determination (calorific value, ash content, moisture) increases. The methods used today are the β^-- and γ-backscattering and γ-ray transmission technique. They are based on the assumption that brown coal is a mixture of two components (mineral matter and hydrocarbons) and the component composition is a constant one. Often, however, this supposition is not correct. An additional demand is the good preparation of coal samples. That is why these methods are not suitable for measurements at the so-called main conveying belt in the open cut. Owing to the availability of correlation functions there is the possibility of determining the coal parameters by analysis of element contents [2]. Fig. 1 and Fig. 2 show that not only the carbon content and the calorific value but the silicon and ash content, too, are correlated.

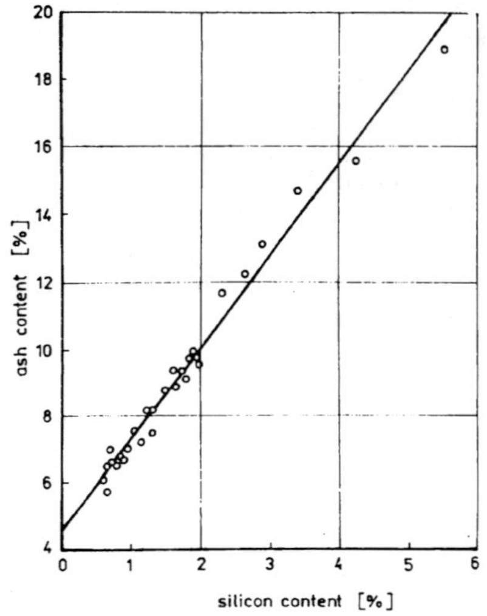

Fig. 1. Correlation between silicon and ash content

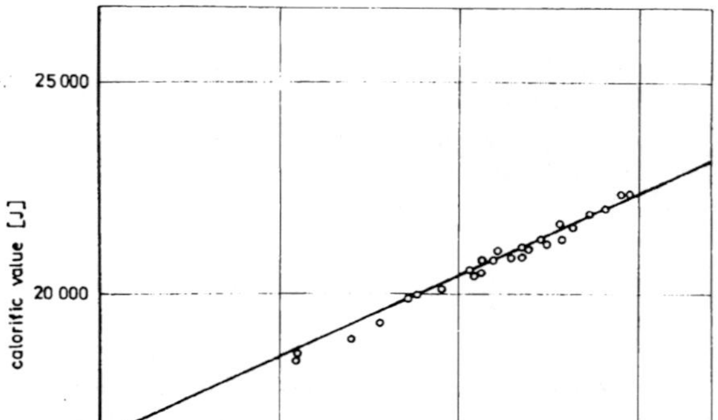

Fig. 2. Correlation between carbon content and calorific value

Due to the great depth of penetration, methods, using γ- and neutron radiation are especially suitable for use in open cuts. The activation analysis with neutrons is a possible method [3] but estimations show that the counting rate in a measuring time interval of approximately 10 minutes (the time for loading a train) is not sufficient. Another method is the use of the inelastic scattering of fast neutrons [4 ... 7]. The cross section of the reaction $^{12}C(n, n'\gamma)^{12}C$ is about 220 mb for neutrons with an energy of 14 MeV. After interaction the ^{12}C nucleus emits a 4.43 MeV γ-radiation at a time interval of less than 10^{-9} s. Therefore the measurement has to be performed concurrently with the time of irradiation with the fast neutrons. As a result of interaction of fast neutrons with the neutron generator, the sample, the shielding materials and the surroundings an intense radiation background with the following reasons is produced:

- γ-radiation as a result of neutron capture reactions (for instance $^1H(n, \gamma)^2H$)
- γ-radiation from the products of neutron activation reactions
- γ-radiation as a result of $(n, n'\gamma)$-reactions with other elements
- interaction of neutrons with the detector material

To decrease this intense background various methods are described:

- time-to-flight-measurements [8]
- pulsed source of neutrons [9]
- extensive shielding [10]

So far, measurement equipment basing on these methods could not be installed in practice. Therefore, new methods were applied at the Dresden University of Technology. The reaction $(n, n'\gamma)$ was coupled with the time correlated associated particle method which has been used previously for exact measurements of fission cross section [11].

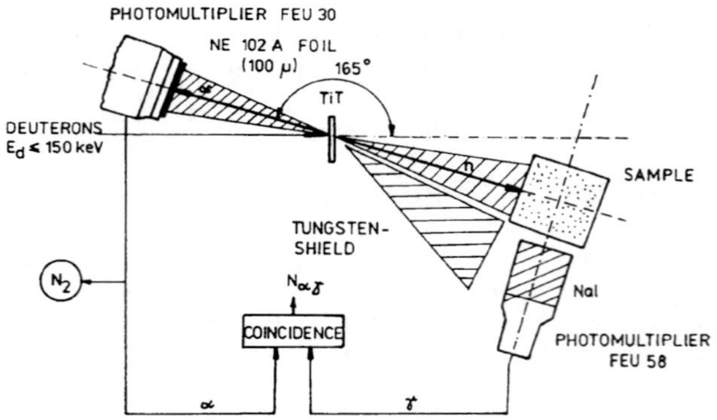

Fig. 3. Principle of $(n, n'\gamma)$ experiment using the time correlated associated particle method in the case of 14.7 MeV neutron energy

3. Principle of Measurements

The basic idea of TCAPM is illustrated in Fig. 3.

A deuteron beam with an energy of about 150 keV hits the target of a neutron generator, where neutrons and associated particles are produced. In the case when the target is a TiT one the associated particles are ^4He-(α-) particles. The associated particle detector is a NE102 A scintillation foil with photomultiplier FEU 30. It counts all charged α-particles within the solid angle $\Delta\Omega_\alpha$ fixed by a diaphragm. The associated particle detector produces the timing signals for associated neutrons within the so-called neutron cone $\Delta\Omega_N$. These neutrons interact with the sample. There are many possible reactions which produce γ-radiation, but only the γ-rays of the (n, n′γ)-reactions are time correlated. Therefore the γ-rays can be detected in coincidence with the associated particles to reduce the troublesome intense noncorrelated background resulting in other reactions and events. The detector is a 3″ × 3″ sodium iodide one with photomultiplier FEU 58. In order to reduce the majority of fast neutrons interacting with the γ-ray detector there is used a tungsten shield.

The number of counts is represented by the formula

$$N = \Phi\eta\varepsilon\nu\sigma mc \frac{N_A}{A_r} (t_2 - t_1)$$

where Φ is the flux of fast neutrons, η the number of γ-quantum per decay, ε the detector efficiency, ν the aboundance of isotopes, c the mass of the sample, σ the cross section, N_A the *Avogadro* number, A_r the relative atomic mass and $(t_2 - t_1)$ the time interval.

4. Block diagram

The block diagram of the electronic equipment is illustrated in Fig. 4.

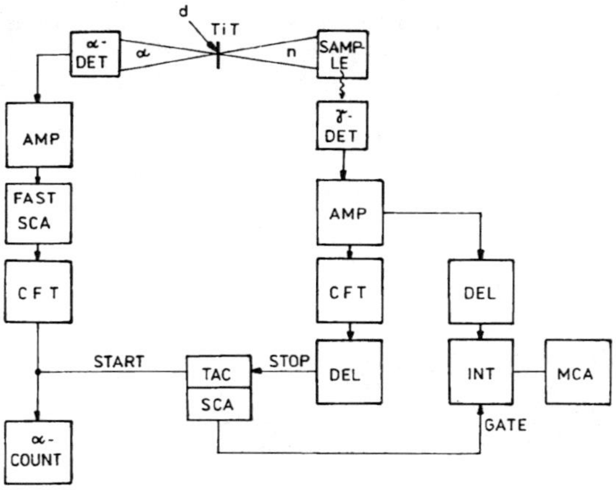

Fig. 4. Block diagram of the electronic equipment
d — deuterons; TiT — TiT-target of the generator; α-alpha cone; n — neutron cone; α-DET — alpha detector; γ-DET — gamma-detector; AMP — fast amplifier; FAST SCA — fast single channel analyzer; CFT — constant fraction trigger; α-COUNT — associated particle counter; TAC — time-to-pulse-height-converter; SCA — single channel analyzer; DEL — delay block; INT — gated integrator; MCA — multi channel analyzer

Due to the necessity of a good separation of α-pulses from pulses of scattered deuterons, γ-rays, tritons and protons a NE 102 A plastic foil scintillation detector is used. Together with a fast analog electronic system it guarantees work at counting rates of about $10^5\ s^{-1}$ and sufficient resolution of about 30% for the α-particles. The pulses from the photomultiplier are amplified by a fast timing amplifier. The selection of α-pulses can be accomplished by using a fast single channel analyzer (timing SCA) consisting of two discriminators, one trigger and a fast coincidence circuit. The constant fraction trigger produces the logic pulses as input signals ("START") for a time-to-pulse-height converter and the associated particle counter. The second input pulse ("STOP") is produced by a similar circuit from the 3″ × 3″ NaI γ-detector to the stop input of the time-to-pulse-height converter (TAC). The spectrometric circuit produces the pulses for a gated integrator. Only the (n, n′γ) processes are time correlated. Therefore it is possible to select the pulses from the TAC so that only the pulses with defined height or a defined time interval between START and STOP will pass the coupled single channel analyzer. Only pulses which are time correlated with the α-particles can pass the integrator and reach the multichannel analyzer.

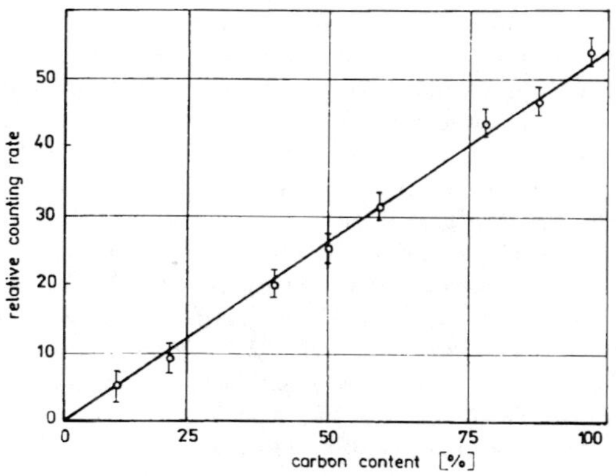

Fig. 5. Calibration curve of ^{12}C

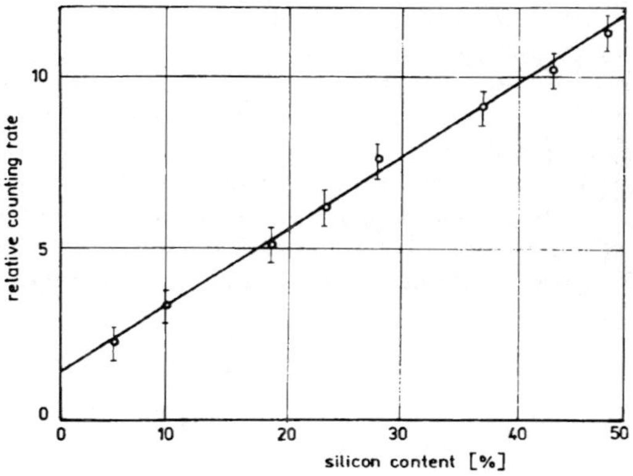

Fig. 6. Calibration curve of ^{28}Si

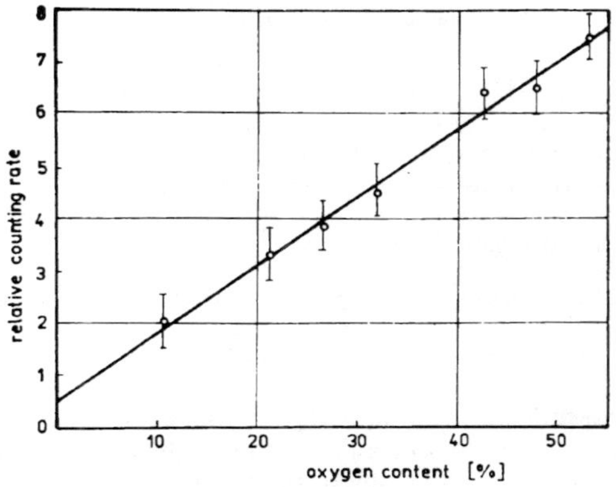

Fig. 7. Calibration curve of ^{16}O

5. Experimental results

The (n, n′γ) measurements using the time correlated associated particle method have been carried out using the *Cookroft-Walton*-Neutron-Generator TNC 9900 (120 ... 150 keV deuteron energy) of the Dresden University of Technology. First measurements were performed without the TCAPM using two different methods: wheter the fore-scatter geometry nor a pulsed generator, too, meet the requirements of brown coal industry for a fast as well as exact measurement system. Therefore a laboratory experiment was developed to determine the carbon, oxygen and silicon content in samples by means of the (n, n′γ)-reaction coupled with the TCAPM. The apex angle of the neutron cone is about 7°, the distance sample-NaI detector is about 10 cm and the distance target-sample is about 32 cm. It was shown that the saturation thickness to determine the carbon content using graphite is about 40 cm. This is in good agreement with MONTE-CARLO calculations. The results of first measurements with a mixture of SiO_2 and graphite are shown in Fig. 5 ... 7. There are represented the relationships between the element contents and the measured counting rates. The statistical error (σ) of the carbon content determination is in the range of 3% for a carbon content of about 50% in the mixture. The measurements and correlation functions show that it will be possible to determine the calorific value and the ash content by (n, n′γ)-reaction coupled with the TCAPM. The use of a neutron generator has still another advantage: the possibility of determination the hydrogen content by neutron capture reaction and silicon and aluminium by fast neutron activation analysis. Such a multielement analysis of brown coal is the base for a precise determination of coal parameters what is necessary for the correct distribution according to the quality of coal, for instance to the briquetting plant, the chemical processing or to the power station.

6. Conclusions

A laboratory measurement system using the (n, n′γ)-reaction coupled with the time correlated associated particle method for determining the element content in coal samples has been developed. Investigations show, that it is possible to find correlation functions between the element contents and coal parameters. First measurements were made with mixtures of graphite and sand. The statistical error for a sample with a carbon content of 50% is about 3%. It is planned to perform test measurements with modified solid angle of neutron cone and Ge(Li) detector.

Received January 31, 1983

References

[1] *B. D. Sowerby*, Nucl. Instr. and Meth. **166** (1979) 371
[2] *L. P. Starchik*, Coke Chem. USSR **5** (1974) 52
[3] *F. Grass*, Nucl. Instr. and Meth. **114** (1974) 49
[4] *T. C. Martin*, Int. Journ. Appl. Rad. **15** (1964) 339
[5] *C. G. Clayton, M. R. Wormald*, IAEA proceedings STI/PUB/598, ISBN 92-0-060082-4, p. 11
[6] *S. W. Jates*, Journ. Rad. Chem. **46** (1978) 343
[7] *F. I. Habiani*, Nucl. Instr. Meth. **134** (1976) 545
[8] *E. M. Diener*, Nucl. Instr. and Meth. **109** (1973) 585
[9] *P. F. McKinlay*, patent FRG 2427127, 1975
[10] *F. C. Engesser*, Journ. Nucl. Ener. **21** (1967) 487
[11] *R. Arlt*, preprint 05-38-79 Dresen University of Technology

Isotopenpraxis **19** (1983) 9, S. 312–314

Betareflectometer SR-77*)

N. S. Batchvarov, I. D. Vankov, L. P. Dimitrov, Ch. D. Shukov, P. A. Pavlov

(Institute of Nuclear Research and Nuclear Energy, Sofia)[1])

A betareflectometer for quantitative determination of heavy and medium heavy elements concentration is described. Important feature of the instrument is the direct read-out of the concentration in percents from the display.

Keywords

alloys; backscattering; beta particles; ores; quantitative chemical analysis

The betareflectometer is designed for quantitative determination of heavy and medium heavy elements in alloys, ores concentrates etc. Measurement is effected by means of the backscattering of beta-particles from samples of the investigated material.

Method

The quantitative analysis method used with apparatus is based on the principle of backscattered beta-particles from the sample.

It is necessary that the atomic number Z of the element under investigation in the sample should differ substantially from that of the matrix (the other elements of the analysed material). For a scatterer thickness greater than so called critical thickness d_{cr}, the flux density of the backscattered beta-particles of a given element does not depend on the thickness of the layer but on the atomic number Z of this element

$$I_r = I_0 K Z^n \quad (1)$$

where I_0 is the flux density of the initial beta-particles,
I_r — the flux density of the scattered beta-particles,
K = 0·002 ÷ 0·42 — coefficient depending on the maximum energy of the spectrum of initial beta-particles,
n = 0·5 ÷ 0·8 depending on the geometry of the experiment.

Formula (1) is valid for pure elements. When the scatterer is a multi-element mixture, the scattered radiation is determined by all elements forming its composition and consequently — by some average characteristics of the scatterer. In such a case we make use of the effective atomic number, i.e. the atomic number of an element if treated under the conditions of the tested sample, would give the same density for the scattered beta-particles flux.

*) Presented at the 2nd Working Meeting "Radioisotope Application and Radiation Processing in Industry", 28th September to 1st October 1982, Leipzig

[1]) Address: Boul. Lenin 72, Sofia, 1184 Bulgaria

When the scatterer consists of only two elements with atomic numbers Z_1 and Z_2 and concentrations of C_1 and C_2, formula (1) becomes as follows:

$$I_r = I_0 K (C_1 Z_1 + C_2 Z_2)^n$$

but since in a bi-component compound in all cases $C_1 + C_2 = 1$, then

$$I_r = I_0 K [Z_1 + (Z_2 - Z_1) C_2]^n \qquad (2)$$

i.e. a singular dependence is obtained between the density of the scattered beta-radiation flux and the concentration of one of the elements.

For determination of concentration of one element in a multielements matrix the sample can be treated as bi-component, replacing Z_1 in eq. (2) by

$$Z_{\text{eff}} = \sum_{i=1}^{r} C_i Z_i \qquad (3)$$

where r is the number of the other elements in the sample. The greater the difference between Z_{eff} and Z_2, the greater the sensitivity of the method.

Principle of operation

An important feature of the instrument is the receiving of the results directly in percents of the element concentration. For this purpose a linear approximation of the equation (2) is accepted which is correct for a limited range of concentration. If the range is larger, it can be separated to several subintervals with the same approximation in each one.

The flux density I_r is measured by a GM-counter, transforming it in a pulse counting rate n. Due to the linear approximation the measured concentration will be

$$C = kn + C_0 \qquad (4)$$

where k is the angular koefficient of the approximating straight line,
C_0 – the intercept.

The counting rate n is measured through the registration of the number of pulses N for a preset time interval T

$$n = N/T \qquad (5)$$

Replacing n from (5) in (4) we get

$$C = \frac{k}{T} N + C_0 \qquad (6)$$

As can be seen, if the time of registration complies

$$T = k \qquad (7)$$

then

$$C = N + C_0 \qquad (8)$$

i. e. the concentration will be equal to the registered number of pulses summed up with the constant C_0. This summation can be easily realized by represetting C_0 in the counter when $C_0 > 0$, or $10^p - C_0$ when $C_0 < 0$ (p is the number of digits in the counter).

For achieving the necessary accuracy the concentration must be measured up to two decimal digits, i.e. the number N must be displayed as XX.XX%. This is equivalent to increasing the number of the registered pulses 10^2 times, which requires the same increase of the measuring time

$$T_0 = 10^2 k \qquad (9)$$

i.e.

$$C = \frac{N}{10^2} + C_0 = \frac{N + 10^2 C_0}{10^2} \qquad (10)$$

The latter equation shows: a) the preset number must be $10^2 C_0$ instead of C_0; b) the final number in the counter have to be divided by 10^2; there are no problems using a decade counter – the decimal point must be placed in the middle of the four digit display and the constant C_0 must be entered with accuracy up to two decimal digits.

The number of registered pulses in each measurement can be calculated from (10)

$$N = 10^2 (C - C_0) \qquad (11)$$

For the normal range of concentration (10 ÷ 100%) N will be $10^3 \div 10^4$ (neglecting C_0) .Such a number of registered pulses can not ensure the required statistical precision – the relative standard deviation is

$$\delta = 1/\sqrt{N} \qquad (12)$$

and for values $\delta \approx 0{\cdot}3 \div 0{\cdot}1\%$ a number of about $10^5 \div 10^6$ pulses is needed. This problem can be solved dividing the input pulses, as well as the pulses from the timer generator by the same coefficient a: if the first term in the right side of eq. (6) would be multiplied and divided by a –

$$C = \frac{ak}{T} \frac{N}{a} + C_0 \qquad (13)$$

the equation remains valid. Therefore the display will show the true concentration value and only the time of measurement increases a times.

Construction

The principle design of the instruments is shown in Fig. 1. The beta-radiation source (1) is hermetically sealed in a Pb-collimator (2), where a conical flux of beta-particles directed towards the sample (3) is obtained. The material

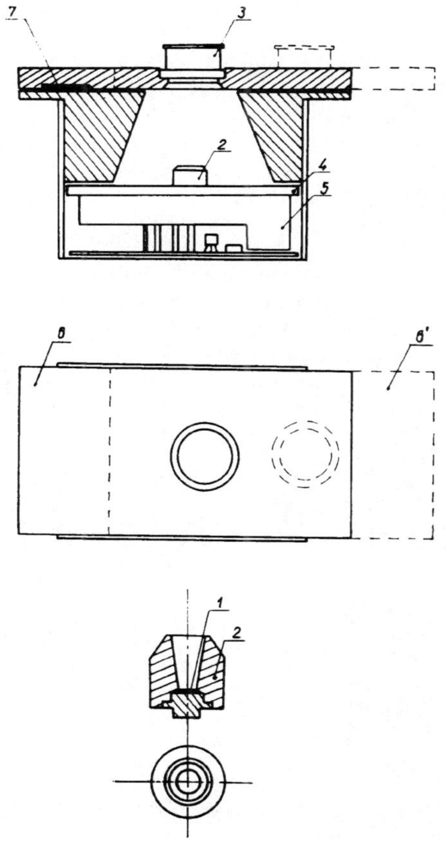

Fig. 1

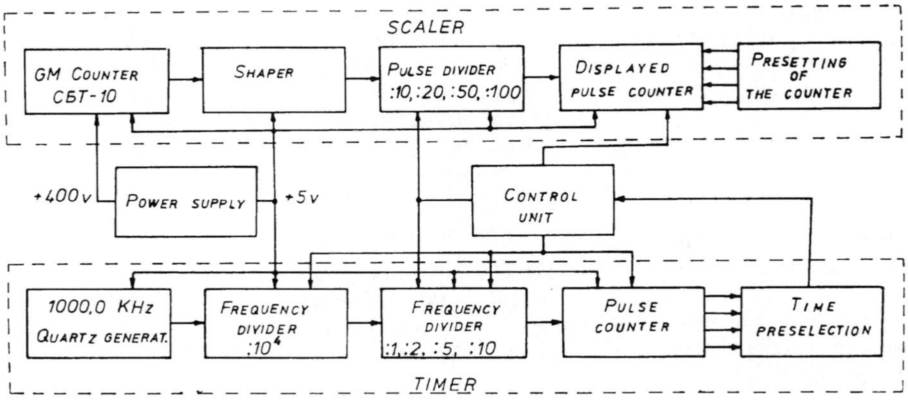

Fig. 2

under investigation is placed in a polyethylene container (3) with a thin mylar film bottom. The beta-radiation scattered by the sample is partially absorbed by an aluminium absorber (4) with the rest reaching the detector (5). The plastic sample holder has two operating positions: "measuring of the sample" (6) and "change of the sample" (6'). In the first position the sample is placed against the radioactive source. In the second position the sample can be changed and in the same time a lead disk (7), placed in the plastic holder, comes above the radioactive source and can be used as a pseudostandard for checking the device. The lead disc closes also the opening of the top panel, thus excluding any possibility of beta-radiation outside the instrument. For security a blocking mechanism does not allow the holder to leave this position if a sample container is not introduced there.

The block diagram of the electronic unit is presented in Fig. 2. It consists of a scaler, a timer, a control unit and the power supply.

The scaler contains a *Geiger-Mueller* counter — type СБТ-10, whose pulses are amplified and formed by the shaper for achieving the TTL compatible voltage level. The formed signals are fed into a frequency divider with a selectable dividing modulo, corresponding to the coefficient a. Its output pulses are applied to a four digit decade counter with a display. The number $10C^2{}_0$ can be preselected by a 4-section commutator.

The timer uses a quartz stabilized generator, the pulses of which after passing through a modulo 10^4 frequency divider are fed into a second adjustable (modulo a) frequency divider. Its output is connected to the input of a four digit decade counter. The time of measurement is preselected by another 4-section commutator choosing the capacity of this counter. The control unit generates the signals necessary for the normal operation of the device. After each signal "start" this unit generates a clear pulse for the timer and another one for writing the preselected number in the pulse counter of the scaler. Permission is then given for the registration of GM-counter pulses in the scaler and the pulses of the quartz generator in the timer. When the preselected time of measurement is up, the output pulse from the timer turns on the control unit thus stopping the measurement.

The instrument's overall view is shown in Fig. 3. There is a special opening on the top panel for placing the sample container in the holder. The basic controls are on the front panel: digital display of concentration, C_0 and "TIME" preselectors, "POWER" on-off switch, push button "START" and coefficient a selectors "x1, x2, x5, x10".

Calibration of the instrument

In order to receive the displayed concentration directly in percents a preliminary calibration have to be done. For this purpose a few sample standards with a 0·1% maximal error have to be prepared from the studied element. Its concentrations must cover the full range of interest. For each sample the number of pulses N_i for a preset time interval t is measured (the 4-section commutator C_0 being set on 00·00). Then the parameters of the approximating straight line can be found by the method of least squares

$$T = \frac{\dfrac{\Sigma n_i \Sigma C_i}{m} - \Sigma n_i C_i}{\dfrac{(\Sigma n_i)^2}{m} - \Sigma n_i^2}, \qquad C_0 = \frac{\Sigma C_i}{m} - T\,\frac{\Sigma n_i}{m}$$

Here $n_i = N_i/t$ and m is the number of measured standard samples.

Basic technical data

1. Minimal quantity of the sample in a container: 2 g.
2. Minimal mean quadratic statistical error: 0·1%.
3. Direct digital read-out of the concentration in %: 0·01% ÷ 99·99%.
4. Digital preselection of measuring time: 0·01 ÷ 999·9 s.
5. Capacity of pulse counter: 10^5, 2×10^5, 5×10^5 and 10^6 pulses.
6. Power supply: AC 110/220 V, 13 VA, 50/60 Hz or DC 12 V.
7. Built-in radioactive source: ^{90}Sr — ^{90}Y, 80 μCi.
8. Dimensions: 220 × 170 × 110 mm³.
9. Weight: 3·5 kg.

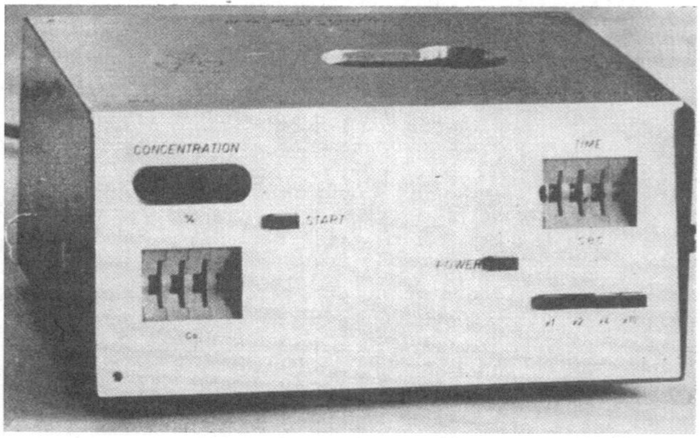

Fig. 3

Received January 14, 1983

Isotopenpraxis 19 (1983) 9, S. 315–317

Kurze Originalmitteilungen

Zum thermischen Verhalten von Uranhydriden, die sich in der isotopen Zusammensetzung der Wasserstoffe unterscheiden

M. Kießling, B. Schossig

(Zentralinstitut für Kernforschung Rossendorf)[1])

Keywords

hydrogen isotopes; isotope effects; isotopic exchange; kinetics; mass spectroscopy; uranium

Organische Verbindungen können mittels verschiedener Wasserstoffisotope positionsspezifisch markiert werden. Durch thermische Zersetzung von Uranhydriden [1,2]-(-deuteriden, -tritiden) ist eine genaue Dosierung der Wasserstoffisotope möglich.

Bei der thermischen Zersetzung von Uranhydriden, die sich in der isotopen Zusammensetzung des Wasserstoffs unterscheiden, treten gleichzeitig Zersetzungs- und Isotopenaustauschreaktionen auf.

Da beispielsweise für spezifische Tritiummarkierungen organischer Verbindungen stets mit $^1H_2$, $^1H^3H$ und ^3He verunreinigtes $^3H_2$-Gas zum Einsatz kommt, besteht aus präparativen Gründen Interesse daran zu wissen, welchen Einfluß die im Unterschuß des Gasgemisches vorkommende Wasserstoffisotopenspezies auf den Markierungsgrad der organischen Verbindung hat.

Die Isotopieeffekte des Wasserstoffs bei der thermischen Zersetzung der Uranhydride müssen dabei ebenso beachtet werden wie die bei der katalytischen Synthesereaktion selbst. In dieser Arbeit sollen unterschiedliche Wasserstoffisotopengemische untersucht werden, die mit pulverförmigem Uran bestimmte Uranhydride bilden. Nach der thermischen Behandlung dieser Uranhydride ändert sich die Gaszusammensetzung, deren Analyse durch massenspektrometrische Bestimmung vorgenommen wurde.

[1]) Anschrift: DDR-8051 Dresden, Postfach 19

Experimenteller Teil

Die verwendeten Gasspezies wurden handelsüblichen Stahlflaschen bzw. Ampullen ($^1H_2$ und $^2H_2$ vom VEB Technische Gase Leipzig; $^1H^2H$ von Stohler Isotope Chemicals Waltham, USA; $^3H_2$ aus der SU) entnommen. Es wurde eine Vakuumapparatur zur Uranhydridbildung und -zersetzung benutzt, wie sie analog zu [3] für die Dehalogenierung von organischen Verbindungen in Anwendung kam. Die verfügbaren Uranspeicher enthalten je 8 g Urangranulat, das durch thermische Aktivierung mit Wasserstoff in pulverförmiges Uranhydrid übergeführt wurde. Nach anschließender thermischer Zersetzung des Uranhydrids und Evakuieren bei 450 °C für 2 Stunden erhielten wir das pulverförmige Uranpulver, das zur darauffolgenden Uranhydridbildung eingesetzt wurde.

Tab. 1. Protium/Deuterium-Verhältnisse im Gasgemisch in Abhängigkeit von der Anzahl der Ausheizzyklen des Uranspeichers

Zyklus	Temperatur d. Uranspeichers °C	Gasdruck Pa	$^1H_2/^2H_2$	$^1H_2/^1H^2H$	$^1H^2H/^2H_2$
Ausgangsgemisch I			3,1	18,2	0,2
1	330	3,00	3,8	1,1	3,3
3	355	4,50	4,2	1,2	3,4
4	355	4,50	5,0	1,2	4,1
5	345	3,92	5,7	1,3	4,4
Ausgangsgemisch II			3,3	0,02	132,4
1	342	3,75	1,2	0,6	2,1
2	355	4,50	1,2	0,6	2,0
3	355	4,50	1,5	0,6	2,3
4	342	3,75	1,9	0,8	2,5
5	355	4,50	1,8	0,8	2,6
6	355	4,50	2,3	0,8	3,0

Tab. 2. Tritium/Protium-Verhältnisse im Gasgemisch in Abhängigkeit von der Anzahl der Ausheizzyklen des Uranspeichers

Zyklus	Temperatur d. Uranspeichers °C	Gasdruck Pa	$^3H_2/^1H_2$	$^3H_2/^1H^3H$	$^1H^3H/^1H_2$
Ausgangsgemisch III			83,3	5,7	14,7
1	187	2,25	51,1	5,4	9,5
2	187	2,25	38,8	4,2	9,4
4	187	2,25	27,9	3,5	7,9
5	187	2,25	26,7	3,4	8,0
6	174	1,88	16,3	2,9	5,7
7	187	2,25	13,7	2,5	5,6
			$^1H_2/^3H_2$	$^1H^3H/^3H_2$	$^1H_2/^1H^3H$
Ausgangsgemisch IV			13,5	1,6	8,3
1	283	3,00	2,2	2,0	1,1
2	283	3,00	3,1	2,8	1,1
3	283	3,00	3,4	2,8	1,2
4	283	3,00	3,6	2,8	1,3
5	283	3,00	3,8	2,8	1,4

Tab. 3. Massenspektrometrisch ermittelte elementare Trennfaktoren

Ausgangs. gemisch	Zyklus	Temp. K	α			Umsetzungsgrad %
			$^1H_2/^1H^2H$	$^1H_2/^2H_2$	$^1H^2H/^2H_2$	
I	5	628	1,11	1,22	1,10	3,3
II	5	623	1,24	1,35	1,09	3,3
			$^1H_2/^1H^3H$	$^1H_2/^3H_2$	$^1H^3H/^3H_2$	
III	8	418	1,14	1,28	1,14	4,9
IV	5	550	1,09	1,13	1,04	3,3

Für die 4 zu untersuchenden Speicher wurden die Ausgangsgemische aus Protium und Deuterium bzw. Protium und Tritiumgas gemischt bzw. den Ampullen entnommen.

Die thermische Zersetzung dieser Uranhydride erfolgte durch langsames Aufheizen des Uranspeichers bis sich, wie in den Tab. 1 und 2 aufgezeigt, ein bestimmter Gasdruck einstellte und eine gute Durchmischung des Gasgemisches in der Vakuumapparatur (einschließlich Probennahmegefäß) gewährleistet war. In Abständen von 0,75 Pa (bis 4,50 Pa) wurden Gasproben entnommen und massenspektrometrisch untersucht. Der 1. Zyklus war somit beendet. Das noch in der Vakuumapparatur befindliche Gas reagierte nun nach Abkühlung reversibel mit dem Uran. Der nächste Ausheizzyklus konnte vorgenommen werden.

Zur massenspektrometrischen Analyse wurde ein Quadrupolmassenspektrometer QMG 101, Fa. Balzers, verwendet.

Andere Parameter, wie die Energie der Stoßelektronen und Geometrie des Einlaßsystems wurden konstant gehalten. Der relative Fehler ermittelter Masseanteile betrug maximal ± 1 Prozent.

Im Verlauf mehrerer Ausheizzyklen stellt sich für Protium/Deuterium ein chemisches Isotopenaustausch-Gleichgewicht ein (Tab. 1).

Für beide Gasgemische läßt sich ableiten, daß im Ergebnis der Zersetzungs- und Isotopenaustauschreaktionen Protium gegenüber Deuterium bevorzugt gebildet wurde, obwohl der Zersetzungsdruck des U^2H_3 gegenüber U^1H_3 das 1,4-fache [4] beträgt.

In Tab. 2 sind die Änderungen der Gaszusammensetzung bei der thermischen Zersetzung der entsprechenden Protium/Tritium-haltigen Uranhydridspeicher ebenfalls in Abhängigkeit von den Ausheizzyklen und vom Gasdruck aufgezeigt. Nach der thermischen Zersetzung des Uranhydridspeichers mit Ausgangsgemisch III beobachteten wir eine erhebliche Verringerung des $^3H_2$-Gehaltes zugunsten der $^1H^3H$-Bildung. Das ist um so bemerkenswerter, als beispielsweise U^3H_3 bei 390 °C einen Zersetzungsdruck von 4,01 Pa und U^1H_3 von 2,48 Pa zeigt.

Die Größe der beobachteten Isotopieeffekte kann durch den elementaren Trennfaktor charakterisiert werden [5].

Die Bestimmung der elementaren Trennfaktoren ist gegeben aus den massenspektrometrischen Gasanalysenwerten in Tab. 3.

Es konnte gezeigt werden, daß die beobachteten Isotopieeffekte unter den gewählten experimentellen Bedingungen von den Masseanteilen der Wasserstoffisotopenspezies des Ausgangsgemisches, von der Temperatur des Uranspeichers bzw. dem sich einstellenden Gasdruck und von der Anzahl der Ausheizzyklen abhängig sind.

Diese Arbeit wurde im Rahmen von präparativen Untersuchungen vorgenommen, bei denen der Einfluß von Isotopieeffekten auf den Markierungsgrad in bestimmten Verbindungen aufgezeigt werden sollte.

Eingegangen am 21. 7. 1982
Überarbeitete Fassung am 26. 1. 1983

Literatur

[1] *T. Hashina, O. Yasuharu,* J. Chem. Phys. **1973,** 855

[2] *E. Wicke, K. Otto,* Z. Physik. Chemie (Neue Folge) **31** (1962) 222

[3] *J. Römer, M. Kießling,* Isotopenpraxis **11** (1975) 150

[4] Gmelins Handbuch der anorganischen Chemie **55** (1970) Erg. Bd. Teil C 1, Berlin: Springer-Verlag

[5] *P. Krumbiegel,* Isotopieeffekte, Wissenschaftliche Taschenbücher WTB, Berlin: Akademie-Verlag 1970

Isotopenpraxis **19** (1983) 9, S. 317–320

Informationen

DDR. Zum Tag des Gesundheitswesens 1982 wurde an die Mitarbeiter der Radiologischen Klinik der Wilhelm-Pieck-Universität Rostock ein Linearbeschleuniger LUE-15 ME für den Einsatz in der Strahlentherapie übergeben. Hersteller ist das Wissenschaftliche Forschungsinstitut für elektrophysikalische Apparaturen ,,NIIEFA", Leningrad. Nach Jena ist Rostock der zweite Standort eines derartigen Gerätes in der DDR.

Mit dem Linearbeschleuniger können Elektronenstrahlenfelder bis zu maximal 20×20 cm² und Energien zwischen 10 und 20 MeV (regelbar in Stufen von etwa 2,5 MeV) sowie Bremsstrahlungsfelder bis zu maximal 35×35 cm² mit einer Grenzenergie von etwa 15 MeV (nicht regelbar) erzeugt werden. Die Strahlenintensität ist für beide Strahlenarten regelbar. Die maximale Dosisleistung in 1 m Target-Abstand beträgt 3 Gy/min. Die Anlage läßt eine Bewegung des Strahlers um den liegenden Patienten von $\pm 115°$ zu, so daß auch eine Bewegungsbestrahlung möglich ist. Dabei beträgt die Drehgeschwindigkeit wahlweise 60 bzw. 30°/min. Die Einstellung des Gerätes und der Bewegungsablauf lassen sich durch einen Programmotor automatisch steuern.

DDR. Vom Staatlichen Amt für Atomsicherheit und Strahlenschutz (SAAS) der DDR sind folgende Berichte erschienen:

SAAS-295 Optimierung des Stichprobenumfangs bei der Kernmaterialkontrolle, Teil 1: Die Stichprobennahme zur Variablenprüfung (Sample size optimization in nuclear material control. Part I. Random sampling for the variables test – *J. Gladitz, K. Hoffmann* u. *K. Willuhn*) 33 S.

SAAS-296 Optimierung des Stichprobenumfangs bei der Kernmaterialkontrolle, Teil 2: Die Stichprobennahme zur Attributprüfung (Sample size optimization in nuclear material control. Part II. Random sampling for the attribute test – *J. Gladitz, K. Hoffmann* u. *K. Willuhn*) 17 S.

SAAS-298 Strahlenschutzprobleme mit umschlossenen Pu-Strahlenquellen (Radiation protection problems with sealed Pu radiation sources – *M. Naumann* u. *Chr. Wels*) 48 S.

SAAS-299 Die berufliche Strahlenbelastung 1979 in der DDR (Occupational radiation exposure in the GDR in 1979 – *K.-Fr. Poulheim, W. Rothe* u. *R. Scheber*) 34 S.

SAAS-300 Die berufliche Strahlenbelastung 1980 in der DDR (Occupational radiation exposure in the GDR in 1980 – *K.-Fr. Poulheim, W. Rothe* u. *R. Scheber*) 25 S.

SAAS-301 Laboratoriumsuntersuchungen der Wirkung von Wärme und Strahlung auf Steinsalz (Laboratory studies of heat and radiation effects on rock salt – *W. Noack, W. Körner* u. *K. Runge*) 47 S.

SAAS-303 Techniques for radiation protection monitoring of personnel at fusion readers (Methoden der Strahlenschutzüberwachung des Personals an Fusionsreaktoren – *in englisch* – *V. Schuricht*) 18 S.

BRD. In Nr. 4, Bd. 41 (1982), der Zeitschrift Atomkernenergie/Kerntechnik wird das Schwerpunktthema Spallationsneutronenquellen behandelt. Folgende Beiträge dazu werden gegeben:

G. S. Baur, The general concept for a spallation neutron source in the FRG, p. 234–242;

P. Cloth, D. Filger u. *R. Hecker,* Nuclear assessment of the DIANE target station, p. 243–252;

J. F. Stelzer, Behaviour of the spallation neutron source target wheel under thermal and mechanical loads, p. 253–258;

H. Stechenmesser, Mechanical design of the target DIANE of the German spallation neutron source project and its rotating target, p. 259–263.

Mit Hilfe der konzipierten Spallationsneutronenquellen sollen thermische Neutronenflüsse um etwa $7 \cdot 10^{14}$ cm^{-2} s^{-1} erreicht werden, und bei einem gepulsten Betrieb kann eine Intensitätsmodulation des Neutronenflusses mit einer Wiederholfrequenz von 100 Hz und einem Spitzenwert von $1{,}3 \cdot 10^{16}$ cm^{-2} s^{-1} erzielt werden (mit einer Pulsdauer von etwa 510 µs). Die Abkürzung DIANE steht für Deutsche intensitätsmodulierte Anlage für Neutronen-Experimente. Im Vergleich dazu hat die IBR II-Anlage im VIK Dubna einen Spitzenneutronenfluß von $2 \cdot 10^{16}$ cm^{-2} s^{-1} und eine Pulsdauer von ungefähr 150 µs; der Hochflußreaktor HFR in Grenoble hat einen Neutronenfluß von 10^{15} cm^{-2} s^{-1}.

Weitere Beiträge in diesem Heft:

K. Starke, Transuranium research in Germany 1939 to 1945, p. 264–266;

Fr. Winterberg, The significance of inertial confinement fusion for fundamental research, p. 267–272

Es handelt sich dabei um thermonukleare Mikroexplosionen, die möglicherweise einmal in der Weltraumfahrt für Antriebszwecke Bedeutung erlangen könnten.

H. Fumoto, E. Zimmer, R. Kiyose and *E. Merz,* Thorium fuel reprocessing with pulse columns (II), p. 273–278.

Es wird die gemeinsame Extraktion und Reextraktion von Th und U im säureunterschüssigen THOREX-Verfahren beschrieben (in Fortsetzung von Teil I, Technol. **58** (1982) 447–464).

BRD. In ,,Topics in Current Chemistry" ist eine zusammenfassende Arbeit von *K. Schwochau* über ,,The analytical chemistry of technetium" mit 155 Literaturhinweisen erschienen. Inorganic Chemistry, Topics in Current Chemistry **96** (1981) S. 109–147 (Springer-Verlag, Berlin–Heidelberg – New York).

Bulgarien. Das Institut für Kernforschung und Kernenergetik der Bulgarischen Akademie der Wissenschaften legte kürzlich erstmalig einen wissenschaftlichen Ergebnisbericht (in Russisch) für das Jahr 1980 vor. Derartige Berichte sollen künftig jährlich erscheinen.

Zu jedem Wissenschaftsbreich werden in Form von Annotationen Angaben gemacht über die Hauptaufgaben, grundlegende Resultate, über die Publikationstätigkeit und weitere Aktivitäten der wissenschaftlichen Mitarbeiter. Dem Heft beigegeben ist eine genaue Übersicht über die innere Struktur des Instituts.

Der Bericht unter dem Titel „Die Tätigkeit des Instituts für Kernforschung und Kernenergetik der Bulgarischen Akademie der Wissenschaften im Jahre 1980" (russ.) kann kostenlos angefordert werden unter folgender Adresse:

Sofia 1184, Boul. Lenin 72
Institute of Nuclear Research and Nuclear Energy.

Frankreich. Im französischen Kernforschungszentrum Cadarache wurde im Oktober 1982 die Anlage Esmeralda eingeweiht, die zur praktischen Erforschung von Natriumbränden und ihre Bekämpfung im Hinblick auf die Sicherheit der natriumgekühlten Schnellen Brutreaktoren dient. Esmeralda umfaßt eine Fläche von 2300 m^2 und wurde in dreieinhalbjähriger Bauzeit errichtet.

Das Commissariat à l'énergie atomique (CEA) verfolgt seit 1972 ein umfangreiches Programm zur theoretischen und angewandten Untersuchung von Natriumfeuern. Dabei wurde eine Reihe von statischen und dynamischen Löschsystemen entwickelt, zum Beispiel Tröge, in denen Natriumbrände ersticken, sowie das Löschpulver Marcalina. Dieses Pulver enthält nichtkorrosive Karbonate der Alkalimetalle und Graphit. Seine Bestandteile reagieren chemisch nicht mit Natrium und bilden beim Kontakt mit diesem keine schädlichen Produkte. Marcalina ermöglicht es, Natriumbrände bei allen Temperatuten zu löschen.

Die Anlage Esmeralda ermöglicht die Durchführung von Natriumbrandexperimenten im Maßstab von Schnellen Brütern der Größe des Superphénix. Zum Beispiel können Flächenfeuer von 200 m^2 mit 70 t Natrium durchexerziert werden, was den pessimistischsten Annahmen über mögliche Unfallabläufe in Brütern entspricht.

(SVA-Bulletin 21/82)

Italien. In Padua (Italien) soll ein wichtiges europäisches Fusionsprojekt realisiert werden, das ursprünglich in Culham (England) hätte gebaut werden sollen. Ausgabenkürzungen der britischen Regierung führen zu einer Verlegung nach Italien. Das Projekt RFX (Reversed-Field Pinch Experiment) wird vom italienischen nationalen Forschungsrat getragen. Mit den Bauarbeitern sollte im Januar 1983 begonnen werden.

Man hofft, daß Japan und/oder die USA sich später noch am Projekt beteiligen werden. Ursprünglich hatten die USA am Projekt mitarbeiten wollen. Der britische Rückzug führte indessen auch zu einem amerikanischen Rückzieher. In den Vereinigten Staaten versucht man nun, in Los Alamos selbst eine Forschungsanlage aufzubauen. Das RFX-Projekt soll für Experimente dienen, die Möglichkeiten einer Vereinfachung der Technologie eines zukünftigen Fusionsreaktors und einer Reduktion der Kosten untersuchen. (SVA-Bulletin 21/82)

Japan. Nach einem Bericht der Financial Times vom 6. Dezember 1982 hat sich Japan möglicherweise eine führende Position bei der Entwicklung von Verfahren zur Massenherstellung von hochkomplexen Elektronik-Chips mit Hilfe von Synchrotronstrahlung erarbeitet. Bei den Mustern, die mit den heutigen photolithographischen Techniken auf Silikonplättchen reproduziert werden, sind Liniendicken von 3 Mikrometern üblich. Der weiteren Miniaturisierung sind Grenzen gesetzt, die u. a. durch die Wellenlänge des bei der Reproduktion verwendeten Lichts bestimmt werden. Feinere Muster mit noch dünneren Linien als bisher könnten auf Silikonplättchen daher erreicht werden, wenn an Stelle von Licht die kurzwellige Synchrotronstrahlung, die von den auf Kreisbahnen zirkulierenden Elektronen in einem Synchrotron ausgesandt wird, verwendet würde.

Die in den Labors der staatlichen japanischen Agency of Industrial Science and Technology hergestellten Muster sollen Linienbreiten von unter 0,6 Mikrometern aufweisen. (SVA-Bulletin 21/82)

UdSSR. Im Heft 6/1982 der Zeitschrift „Atomnaja Energia", Jahrgang 53, sind u. a. folgende Beiträge in russisch enthalten:

Atomic Science and Technology in the USSR (*Petrosyants, A. M.*);
A Comparison of Technical and Economic Characteristics of Modern LWR and LMFBR Power Plants (*A. A. Rineiski*);
The Experimental Basis of a New Method for Monitoring Radioactive Aerosols Released from RBMK Atomic Power Stations (*Moskvin, L. N., Leontyev, G. G., Nekrestyanov, S. N., Eperin, A. P., Shcherbina, V. G., Mokhnachyov, A. G.*);
Gas Desorption from Metals by Molybdenum-Ion Irradiation (*Katrich, N. P., Kanishchev, V. N.*);
Simulation of Reactor Conditions in the Investigation of Radiation Destruction of Material Surfaces (*Kalin, B. A.*);
Analysis and Evaluation of Neutron Cross Sections for the Curium Isotopes (*Belanova, T.*);
Neutron Transmission Function for the Unresolved Resonance Region (*Komarov, A. V., Lukyanov, A. A.*);
Currents Generated in Pyroelectrics During Gamma Irradiation (*Levin, B. A.*).

IAEA. In Nr. 4, Vol. 24 (1982), des International Atomic Energy Agency Bulletin sind folgende Beiträge enthalten:

Summer 1942 in Chicago, by *B. Goldschmidt*;
Safeguards and nuclear safety: a personal perspective, by *L. Manning Muntzing*;
Nuclear power experience, by *J. Daglish*;
Nuclear power and technology transfer, by *H. J. Laue*;
Energy investment in developing countries, by *Y. Rovani*;
Is the energy debate really about energy? by *B. Wolfe*;
Radiation risk assessment: the 1982 UNSCEAR report.
Conference reports:
Radioimmunoassay for medical diagnosis and research, by *R. A. Dudley* and *B. Vavrejn*;
Mineral exploration, mining and processing, by *Yu. G. Sevastyanov*;
Quality control for nuclear medical instruments, by *E. U. Buddemeyer*;
Plasma physics and controlled nuclear fusion, by *M. Leiser*.

DDR. Aus der Schriftenreihe „Anwendung von Isotopen und Kernstrahlungen in Wissenschaft und Technik" der Isocommerz GmbH, Berlin, ist die Nummer 13 „^{15}N

Experiments in Fertilization Research" (*P. Krumbiegel* u. *H. Bornhak*) erschienen. Ferner können die neuen Kataloge „Reference Sources" und „Test- und Referenz-Quellen" angefordert werden bei Isocommerz GmbH, DDR-1115 Berlin, Lindenberger Weg 70.

Vom Zentralinstitut für Kernforschung (ZfK) der Akademie der Wissenschaften der DDR, 8051 Dresden-Rossendorf, ist eine Zusammenstellung über „Radiopharmakologie der Myocardagentien" (*R. Münze*) mit 346 Literaturzitaten erschienen.

BRD. In Nr. 2, Februar 1983, der Zeitschrift Atomwirtschaft/Atomtechnik wird als Schwerpunktthema die Wiederaufbereitung von Kernbrennstoffen behandelt. Folgende Beiträge dazu werden gegeben:

W. Stoll: Argumente zur Schließung des Brennstoffkreislaufs durch Wiederaufbereitung;
W. Schüller: Betriebserfahrung mit der Wiederaufbereitung — eine internationale Übersicht;
N. Pelzer: Endbeseitigung radioaktiver Abfälle als Staatsaufgabe — ein internationaler Rechtsvergleich;
K. Wasinger: Zwischenlager für abgebrannte Brennelemente im Kernkraftwerk Atucha I (CNA);
H.-J. Laue: Probleme des Transfers nuklearer Technologie — ein Tagungsbericht;
Ch. Stubbe: Rost in Grafenrheinfeld? — eine Kernenergie-Kontroverse;
K. Becker: Zum gegenwärtigen Stand der kerntechnischen Normung;
J. Wolters: LWR-Sicherheitsforschung — neue spektakuläre Ergebnisse kaum noch zu erwarten — ein Tagungsbericht;
L. F. Franzen: Stand der Risikountersuchungen bei Kernkraftwerken — ein Tagungsbericht.

Schweiz. Ein provisorischer gesamtschweizerischer Durchschnittswert der Radonkonzentration in Wohnräumen von 1,5 pCi/L, entsprechend einer mittleren effektiven Äquivalentdosis von 120 mrem/Jahr, wurde im kürzlich veröffentlichten Bericht „Radon in Wohräumen in der Schweiz — Ergebnisse der Vorstudie 1981/82" vom Eidg. Institut für Reaktorforschung (EIR) in Würenlingen ermittelt. Dieser Wert ist direkt zu vergleichen mit der übrigen natürlichen Strahlenbelastung von rund 127 mrem/Jahr. Radon ist damit die größte einzelne Komponente der natürlichen Strahlenbelastung in der Schweiz. Das in der Atemluft vorhandene natürliche radioaktive Edelgas Radon stammt aus dem radioaktiven Zerfall von natürlicherweise im Erdboden und in Baustoffen vorhandenen Stoffen wie Uran und Thorium. Im Winterhalbjahr 1981/82 wurden in der Schweiz in 123 Wohnhäusern in ausgewählten Regionen je drei passive Radondosimeter exponiert und im Sommer 82 ausgewertet. Die Dosimeter wurden meist im Keller, im Wohnzimmer und in einem Schlafraum plaziert. Die mittleren Radonkonzentrationen in den ausgemessenen Räumen lagen zwischen 0,4 und 105 pCi/L (0,015 — 3,9 kBq/m^3) mit einer starken Häufung zwischen 0,5 und 2 pCi/L (20 — 70 Bq/m^3). Starke regionale Unterschiede ergeben sich als Folge der verschiedenen geologischen Formationen. Auch der Stockwerkeinfluß ist zu erkennen. Das Radon gelangt zur Hauptsache aus dem Erdboden in die Häuser, nur in geringerem Maße aus Baumaterial. Lokale Unterschiede können sehr groß sein; einige überraschende Fälle wurden näher untersucht.

Als Anschlußprogramm ist gegenwärtig die Untersuchung der Einflüsse von Energiesparmaßnahmen (Abdichten von Fenstern) auf die Radonkonzentration in ausgewählten Häusern in drei Regionen im Gange. (SVA-Bulletin 3/83)

Frankreich. Im Auftrag der Electricité de France (EDF) hat die Novatome kürzlich mit den Arbeiten für die Entwürfe für das nukleare Dampferzeugungssystem begonnen, mit dem die für die Zukunft geplanten 1500-MW-Brüterkraftwerke Frankreichs ausgestattet werden sollen. Ein detailliertes Projekt könnte den französischen Behörden bis 1985 vorgelegt werden. Damit wäre der Weg für einen Entscheid über den Baubeginn geebnet, der 1986 gefällt werden könnte. Bis dahin werden als wichtige weitere Entscheidungsgrundlage erste Betriebserfahrungen mit dem 1200-MW-Demonstrationskraftwerk Superphénix, das den Betrieb 1984 aufnehmen soll, vorliegen. (SVA-Bulletin 2/83)

Frankreich. Am 3. Januar 1983 nahm das französische Schnellbrüter-Prototypkraftwerk Phénix (250 MW) in Marcoule den Betrieb mit allen drei Kreisläufen wieder auf, nachdem es am 16. Dezember 1982 wegen eines Lecks im Dampferzeuger Nr. 1 abgestellt werden mußte. Auf Grund der Erfahrungen, die anläßlich eines ähnlichen Zwischenfalls im April 1982 gesammelt worden waren, konnte die Natrium-Wasser-Reaktion äußerst rasch unterbrochen und die Beeinträchtigung der Anlage minimal gehalten werden. Wie das Commissariat à l'énergie atomique (CEA) mitteilte, arbeitete Phénix bei Eintreten des Zwischenfalls mit Volleistung. Das Leck trat im nicht-nuklearen Teil des Kraftwerks auf und betraf das nicht-radioaktive Natrium des Sekundärkreislaufs. (SVA-Bulletin 2/83)

Frankreich. Der Stromverbrauch stieg 1982 in Frankreich nur leicht von 258,3 Mia kWh auf 261 Mia kWh, wie kürzlich veröffentlichten Angaben der Electricité de France zu entnehmen ist. Zur Stromproduktion trug die Kernenergie 38,7% bei. Diese Elektrizität wurde von 32 Blöcken an 12 Standorten geliefert: Marcoule G 3; Chinon-A2, -A3, -B1; St. Laurent-A1, A2, -B1, -B2; Monts d'Arrée; Phénix Marcoule; Chooz-A; Fessenheim-1, -2; Bugey-1 bis -5; Tricastin-1 bis -4; Gravelines-B1 bis -B4; Dampierre-1 bis -4; Blayais-1, -2.

1982 wurden nur zwei Nuklearblöcke neu ans Netz geschaltet, nämlich Blayais-2 am 17. Juli und Chinon-B1 am 30. November. Im Vergleich mit den sieben Einheiten, die 1980 und den acht Einheiten, die 1981 ans Netz angeschlossen wurden, scheint dies eine eher bescheidene Leistung zu sein. Dazu ist allerdings zu bemerken, daß 1980 und 1981 zwei außergewöhnliche Jahre gewesen sind. 1983 sollten vier oder fünf Einheiten der 900-MW-Klasse (Chinon-B2, Blayais-3 und -4, Cruas-1 und -2) und der erste Block der 1300-MW-Klasse (Pluel-1) in Betrieb genommen werden. (SVA-Bulletin 3/83)

Belgien. Die Stromproduktion des Kernkraftwerks Doel und der belgischen Anteile an den Kernkraftwerken Tihange und Chooz betrug 1982 nach ersten Schätzungen 12711 Mio kWh und damit rund 27,7% der belgischen Elektrizitätsproduktion. Doel-1 (390 MW) produzierte 3185 Mio kWh und erreichte eine außergewöhnliche Arbeitsausnutzung von 92,7%. Die zweite Einheit von Doel (390 MW) konnte eine Arbeitsausnutzung von 75,2% erreichen und produzierte 2582 Mio kWh. Die erste Druckwasserreaktoreinheit (880 MW) des Kernkraftwerkes Tihange produzierte 1982 6171 Mio kWh mit einer Arbeitsausnutzung von 81%. (SVA-Bulletin 3/83)

Großbritannien. Die erste Einheit des an der Südküste Englands gelegenen Kernkraftwerkes Dungeness-B wurde im Dezember 1982 erstmals kritisch. Dungeness-El ist mit einem fortgeschrittenen gasgekühlten Reaktor (ACR) ausgerüstet und hat eine Leistung von 600 MW. Man rechnet damit, daß im Laufe dieses Jahres auch die zweite Einheit des Kernkraftwerkes in Betrieb genommen werden kann. Das Happy End einer eher unglücklichen Geschichte ist somit nahe. Verschiedenste Verzögerungen sorgten für eine ungewöhnlich lange Bauzeit: Die Arbeiten hatten 1966 begonnen. Weitere AGR-Einheiten sollen 1983 in Hartlepool und Heysham-1 ans Netz angeschlossen werden.

(SVA-Bulletin 3/83)

Finnland/Frankreich/Schweden. 40,3% der gesamten im Jahre 1982 in Finnland erzeugten Elektrizität, nämlich 15,92 Mia kWh, stammten aus Kernkraftwerken. Damit steht Finnland in puncto Atomstromanteil an der Spitze aller Länder der Welt. An zweiter Stelle folgt Frankreich mit 38,7%. Die französischen Kernkraftwerke produzierten 1982 über 100 Mia kWh, was beinahe dem dreifachen Elektrizitätskonsum der Schweiz im Jahre 1981 entspricht. Obschon die definitiven Zahlen noch nicht vorliegen, dürfte Schweden bezüglich Atomstromanteil 1982 hinter Finnland und Frankreich den dritten Rang einnehmen.

(SVA-Bulletin 3/83)

USA. Über Three Mile Island, TMI, den größten Störfall in der Geschichte der kommerziellen Nutzung der Kernenergie, ist es zu einer gütlichen Einigung gekommen: Die Besitzerin des Kernkraftwerkes, die General Public Utilities (GPU), hat die gerichtlichen Schritte gegen die Herstellerfirma Babcock und Wilcox eingestellt. Wie am 24. 1. 1983 bekannt gegeben wurde, verzichtete die GPU auf den geforderten Schadenersatz. Die Gesellschaft wird jetzt in Form von Rückerstattungen für Ausrüstungen und Dienstleistungen in den kommenden zehn Jahren $ 37 Mio von Babcock und Wilcox erhalten. (SVA-Bulletin 3/83)

Isotopenpraxis 19 (1983) 9, S. 320–328

Buchbesprechungen

Umwelt-Stress
Tagungsbericht der III. Arbeitstagung
Umweltbiophysik vom 25.–29. 3. 1981, Templin

Hrsg.: K. Unger, J. Schuh. Martin-Luther-Universität Halle-Wittenberg, Wissenschaftl. Beitr. 1982/35 (P17). Halle (Saale) 1982. 387 S., 134 Abb., 21 Tab.; kart. M 24,50

Die III. Arbeitstagung Umweltbiophysik wurde vom 25. bis 29. 3. 1981 in Templin von der Gesellschaft für physikalische und mathematische Biologie der DDR, gemeinsam mit der Biologischen Gesellschaft der DDR unter Mitwirkung der WGMA und der Martin-Luther-Universität Halle-Wittenberg, veranstaltet. Der vorliegende Tagungsbericht ist eine Auswahl der auf dieser Arbeitstagung gehaltenen Vorträge bzw. Poster – insgesamt wurden 72 Arbeiten erfaßt. Umwelt-Stress soll in dem hier gemeinten Sinne alle psychischen und physischen Reaktionen von Organismen bedeuten, die als Folge von Umweltfaktoren (physikalische, chemische, meteorologische u. a.) auftreten.

Der Tagungsbericht ist in 3 große Teilgebiete gegliedert: Umwelt – Stress – Mensch – Tier (31 Arbeiten, S. 1–151), Umwelt – Stress – Pflanze (31 Arbeiten, S. 152–321) und Umwelt – Stress – Ökosystem (10 Arbeiten, S. 322–387).

Luftverunreinigungen zu analysieren, ihre Folgen auf Mensch, Tier und Pflanze zu studieren und entsprechende Maßnahmen administrativer Art daraus abzuleiten, sind uns eine längst gewohnte Folge zwangsläufiger Schritte. Luftverunreinigungen können auch als chemische Stressoren in den Rahmen der anderen auf uns wirkenden Stressoren gestellt werden, um so das wirkende Wechselverhältnis besser erkennen zu können – neue Begriffe tauchen auf wie etwa Psychochemie („The Psychochemistry of Pollutants" nach *R. W. Russel*), neue Wechselbeziehungen wie etwa zwischen Toxikologie und Biophysik, Umwelttoxikologie und Umweltbiophysik u. ä., worauf *K.-H. Lohs* (Leipzig) aufmerksam machte. Die oft erahnten Zusammenhänge zwischen Wetter und verschiedenen Krankheitsverläufen einschließlich der Sterblichkeit waren Gegenstand der Untersuchungen von *Chr. Haase* und *E. Turowski* (Berlin), die solche Zusammenhänge deutlich machten und eine Ausgliederung von Landschaftsklimaten in der DDR zeigten. Daß eine Reihe von Vorträgen dem Stressfaktor SO_2 für Pflanzen gewidmet waren, war zu erwarten. Erstaunlich ist, aus berufenem Munde zu hören, daß in Tharandt auf eine über 100 Jahre alte Tradition an Untersuchungen der Rauchschäden auf unsere Wälder zurückgeblickt werden kann – die ersten Untersuchungen wurden bereits 1840 von *Stöckhardt* vorgenommen –, welch eine lange Reihe von Jahren, in der das Wissen um die Schäden angehäuft wurde. Die Tharandter Schule hat vor allem über komplexe Stressfaktoren berichtet: SO_2 plus Frost-Stress, SO_2 plus Dürre-Stress, was am Prolingehalt von jungen Koniferen gezeigt werden konnte, *M. Tesche* (Tharandt). Die günstigen Wirkungen von kleinen Dosen ionisierender Strahlung auf das Saatgut von Maispflanzen konnte *R. Koepp* (Berlin) deutlich machen.

Im Beitrag von *H. Lieth* und *G. Esser* (Osnabrück) wurde ein verbessertes Modell für die Nettoprimärproduktion im globalen Maßstab und für die BRD vorgestellt. Die bisherigen Modelle (Miami-Modell, Montreal-Modell, Manaus-Modell, The Hague-Modell), die bereits Eingang in die Lehrbücher der Botanik gefunden haben, wurden durch die Einbeziehung der Böden und Bodenqualitäten entscheidend verbessert; und der Tradition folgend, die neuen Modelle der Autoren immer mit dem Ortsnamen zu benennen, wo sie zum erstenmal vorgestellt wurden, heißt nunmehr dieses Modell Templin-Modell und dürfte künftig öfters zu hören und zu lesen sein; ein schönes Beispiel für eine erfolgreiche Arbeitstagung.

Tagungsberichte (Proceedings) zeichnen sich immer durch eine besonders hohe Aktualität nicht nur der dargebotenen Themen, sondern auch der zitierten Literatur zu den Themen aus. So auch der vorliegende Tagungsbericht. Es sind insgesamt 567 Arbeiten zitiert, wovon 60,1% aus den Jahren 1976–82 stammen, das Maximum liegt bei 11,5% aus dem Jahre 1978, und allein das Jahr 1981 ist mit 48 Zitaten erfaßt. Leider haben nicht alle Autoren ihre Arbeiten mit

Literaturzitaten belegt — insgesamt 10 Arbeiten sind ohne Literaturhinweise, und bei 2 Arbeiten kann die Literatur bei den Autoren angefordert werden, ein Verfahren, das man grundsätzlich vermeiden sollte, denn ein Besitzer dieses Tagungsbandes möchte nicht erst in briefliche Verbindungen eintreten, um seinen Band vollständig zu haben. Überhaupt mangelt es sehr an Einheitlichkeit der Literaturhinweise, und manche Abbildung ist durch die Verkleinerung beim Offsetdruck kaum noch zu lesen. Ein Teilnehmerverzeichnis und ein Autorenverzeichnis hätten zur Abrundung des Ganzen beigetragen. Verwunderlich ist, daß kein Beitrag über unsere stressige Industriegesellschaft, insbesondere aus dem Bereich der Großstädte mit ihren eigenen Stressoren, vorgetragen wurde. Insgesamt aber eine wertvolle Sammlung von Ergebnissen, die einen Überblick über moderne Umweltforschung geben und die Kompliziertheit der Probleme erkennen lassen.

H. Hübner

Jahrbuch der Atomwirtschaft 1982

Hrsg.: W. D. Müller, R. Hossner. Düsseldorf: Handelsblatt GmbH. 458 S. DIN A4; DM 52,—

Die Bedeutung der Kernenergie für die wirtschaftliche Entwicklung der Länder nimmt infolge der immer sichtbarer werdenden Verknappung der fossilen Brennstoffe ständig zu. Auch wenn die Lösung einer Reihe von Fragen der Wiederaufbereitung der Brennstoffe und der Entsorgung der kernenergetischen Anlagen noch umfassender Forschungen und technologischer Entwicklungen bedürfen, sind die bereits heute international erzielten Ergebnisse der friedlichen Nutzung der Kernenergie, insbesondere der Kernenergetik, bemerkenswert. Die Produktion dieser Anlagen hat immer größeren Anteil an den energetischen Bilanzen der Länder. Das vorliegende Buch ist eine gut systematisierte und um zusätzliches Material ergänzte Informationssammlung aus dem Jahrgang 1981 der Zeitschrift „atomwirtschaft/atomtechnik". Es behandelt den Stand und die Entwicklungstendenzen der Kernenergie in der Welt und gliedert sich in vier Teile: Perspektiven, Informationen, Daten und Register.

Das Kapitel „Perspektiven" behandelt vorwiegend die Probleme der nuklearen Entsorgung, Kernbrennstoffversorgung und des Atomenergierechts. Es werden die in der BRD und in Westeuropa geführten Diskussionen zu Notwendigkeit, Effektivität und Sicherheit der kernenergetischen Anlagen zusammengefaßt. In einem gesonderten Kapitel charakterisiert *L. Wiesner* Stand unf Perspektiven der industriellen Anwendung von Radionukliden und strahlentechnischen Verfahren.

Das Kapitel „Information" enthält die nach Ländern und Sachgebieten geordneten Kurznachrichten zu verschiedenen Aspekten der Forschungspolitik, der Planung und Finanzierung der Kernforschung und Kerntechnik in den einzelnen Ländern der Welt. Diese Information basiert im wesentlichen auf analogen Informationen der einzelnen Hefte der Zeitschrift „atomwirtschaft/atomtechnik".

Zu den anderen Teilen des Buches gehört eine Sammlung von Daten und Fakten über die Kernkraftwerke und Kernreaktoren, über die Bedeutung der Kernenergie in der Kernwirtschaft der Länder, Betriebsergebnisse der Kernkraftwerke, wissenschaftliche Daten zur Kernenergieentwicklung sowie faktografische Angaben über öffentliche Ausgaben und Haushalt für die Kernenergie der BRD, Kernbrennstoffzyklus, Nutzung von Radionukliden sowie Sicherung und Umwelt. In diesem fast 70 Seiten umfassenden Teil des Buches wird eine Reihe von Angaben in Tabellen gesammelt und z. T. grafisch dargestellt. Sie geben einen guten Aufschluß über die Entwicklung der Kernenergetik, ihre Effektivität und Ökonomie im Vergleich mit anderen Arten der Energiegewinnung und gestatten, neben einer umfassenden Analyse des Standes der Kernenergieentwicklung in der Welt, Aussagen über die Kosten und den Nutzen der Kernenergie.

Ein gut ausgearbeitetes Registerwerk (Sach- und Namenregister) gibt dem Leser die Möglichkeit, schnell an die benötigte Information heranzukommen und sich mühelos in dem komplizierten Stoff des Buches zu orientieren.

Das Buch kann deshalb zur Einschätzung bestimmter Entwicklungstendenzen der Kernenergie auf verschiedenen Ebenen der Leitungstätigkeit der Forschungsprozesse verwendet werden.

L. Jankowski

Statistik und Versuchsplanung in Technik und Wissenschaft

N. L. Johnson, F. C. Leone; Moskau; Verlag Mir 1981. 516 S., 44 Abb., 153 Tab. (russisch), Rbl 2,60; M 13,—

Das vorliegende Buch mit dem Untertitel „Methoden der Versuchsplanung" ist der zweite Band der Monographie „Statistik und Versuchsplanung in Technik und Wissenschaft" der beiden bekannten amerikanischen Autoren. Im ersten Band, der 1980 in russischer Sprache im Verlag Mir erschien, sind die Grundlagen der Wahrscheinlichkeitsrechnung, Methoden zur Prüfung von Hypothesen, die Theorie statistischer Entscheidungen und die Regressions- bzw. Korrelationsanalyse enthalten. Aufbauend auf diesen Grundlagen umfaßt der zweite Band die Varianz-, Sequential- und Diskriminanzanalyse. Den weitaus größten Teil (etwa die Hälfte des Buches) nimmt hierbei die Varianzanalyse ein, für die ausführlich verschiedene Versuchspläne (einfache und modifizierte Pläne, Pläne höherer Ordnung, Faktorpläne) dargelegt werden. Das Ziel des Buches ist nicht nur in der Vermittlung der angegebenen statistischen Methoden zu sehen, sondern es besteht vor allem darin, dem Leser zu zeigen, wie diese Methoden richtig anzuwenden sind. Die gewählte mathematische Strenge dient damit neben der Exaktheit in den Formulierungen und der Eleganz in den Beweisen vor allem dem Schutz vor falschen Schlußfolgerungen. Leser, die über Kenntnisse auf dem Gebiet der Wahrscheinlichkeitsrechnung und mathematischen Statistik verfügen, können den vorliegenden zweiten Band unabhängig vom ersten Band verwenden, obwohl die Autoren gelegentlich auf diesen verweisen.

Nützliche Informationen kann aber auch derjenige Leser erhalten, der mit den Grundlagen der Wahrscheinlichkeitsrechnung und mathematischen Statistik nicht so vertraut ist. Die angegebenen Beweise kann man übergehen und sich auf die Aufgabenstellung und Lösung bzw. Schlußfolgerungen konzentrieren. Die 78 durchgerechneten Beispiele sowie die ca. 170 Übungen, die einen unterschiedlichen Schwierigkeitsgrad haben und zu denen im Anhang fast immer ein ausführlicher Lösungsweg angegeben ist, kommen diesem Leserkreis entgegen. Die vielen Beispiele und Übungen können einen oberflächlichen Leser dazu verleiten, das Buch als einen Sammelband fertiger Rezepte zu betrachten. Ein genaueres Hinsehen überzeugt jedoch davon, daß die aus verschiedenen Gebieten der Wissenschaft und Technik ausgewählten Beispiele und Übungen dazu dienen, die Probleme exakt zu erfassen, Aspekte zu verdeutlichen und die Ergebnisse richtig zu interpretieren. Einzelne Kapitel (insbesondere im Abschnitt Sequentialanalyse) können beim ersten Durcharbeiten übergangen werden, da sie der Behandlung spezieller Fragen gewidmet sind oder einen anderen Lösungsweg beschreiben.

Das Buch stellt für Spezialisten, die sich mit der Anwendung von Methoden der mathematischen Statistik und der Versuchsplanung befassen, ein wertvolles Arbeitsmittel dar und ist für Studenten höherer Studienjahre eine fundierte und empfehlenswerte Grundlage für die angegebene Problematik.

G. Kummer

Wörterbuch der Mikroelektronik
Englisch – Deutsch/Deutsch – Englisch

Hrsg.: IWT Verlag GmbH, Vaterstetten. Alleinvertieb: VDI-Verlag. 1980. 288 S.; kart. DM 44,–

Für die Mikroelektronik, die in allen Bereichen der Wirtschaft, begonnen bei der Forschung und Entwicklung bis hin zur Produktion, Medizin usw. oder der staatlichen Planung und Verwaltung, festen Fuß gefaßt hat, ist fast eine eigene Fachsprache entstanden, die nahezu ausschließlich aus englischem Vokabular besteht. Bei der Verbreitung und Bedeutung dieser neuen Technologie ist eine eindeutige Verständigung auf diesem sich schnell entwickelnden Fachgebiet wichtig.

Das vorliegende Wörterbuch versucht mit etwa 7000 einschlägigen Begriffen in Englisch – Deutsch und Deutsch – Englisch dem entgegenzukommen. Es zeigt die Verwendung der Fachwörter, wie sie heute bei Ingenieuren, Technikern, Wissenschaftlern und Lehrern gebräuchlich sind.

Das Wörterbuch der Mikroelektronik ist jedem auf diesem Fachgebiet Tätigen sehr zu empfehlen.

J. Holzhey

Induced Mutations – A Tool in Plant Research
Proc. of a Symp., Vienna, 9 – 13 March 1981

Vienna: IAEA 1981, 538 S., div. Abb. u. Tab. (Series – STI/PUB/581); ÖS 840.–

Dieser Tagungsband der IAEA enthält den Bericht eines internationalen Symposiums über die induzierten Mutationen als Mittel zur Verbesserung von Nutzpflanzen. Die Tagung wurde gemeinsam mit der FAO, der Lebensmittel- und Landwirtschaftsorganisation der Vereinten Nationen organisiert.

Der Band enthält Vorträge verschiedener Sektionen, in denen induzierte Mutationen in Beziehung gesetzt werden zu den Fortschritten der Genetik, der pflanzlichen Evolution, der Pflanzenphysiologie, Parasit-Wirtspflanzen-Beziehungen, pflanzlichen Symbiosen, in vitro-Kulturtechniken, Genökologie und Pflanzenzüchtung. Nahezu alle Abschnitte werden eingeleitet durch einen Artikel mit Übersichtscharakter, wobei keine Beschränkung der Themenbehandlung auf strahleninduzierte Mutationen erfolgt. Hierdurch liefert das Buch einen guten Überblick über den internationalen Stand sowie durch insgesamt 34 Artikel und 18 Posterberichte auch zahlreiche Detailergebnisse dieses Arbeitsgebietes.

Strategien zur Verbesserung der Eigenschaften und Erträge von Nutzpflanzen, ohne vorhandene ökologische Anpassungen zu beeinträchtigen (*J. N. Rutger* und *M. L. Peterson*; *W. Gottschalk*), werden ebenso behandelt wie die Möglichkeiten, Mutanten mit besserer Anpassung an veränderte Bedingungen, z. B. größere Kältetoleranz (*F. Walther* und *W. Preil*), oder solche mit verändertem Gehalt an Inhaltsstoffen (*K. W. Pakendorf* und *P. J. van Rensburg*) zu erhalten.

Besonders die Artikel des ersten Abschnitts geben einen Einblick in die Möglichkeiten, die sich durch Nutzung induzierter Mutationen für genetische Untersuchungen eröffnen. Identifikation der Genorte, Zusammenspiel mehrerer Gene bei der Genexpression (z. B. Positionseffekte), Möglichkeiten des Genengineering, Probleme der Selbstinkompatibilität und der physiologischen und biochemischen Genetik (*J. MacKey*) werden ebenso behandelt wie die Ergebnisse langjähriger Mutationsexperimente an Gerste (*F. Scholz* und *G. Künzel*) oder die experimentellen Möglichkeiten, die der Einsatz ionisierender Strahlung zur Induktion parthenogenetisch diploider Samen eröffnet (*K. K. Pandey*). Besonders verwiesen sei auf die Artikel zum Einsatz der in vitro-Kulturtechnik (Sektion 6), in welche die moderne Züchtungsforschung große Erwartungen setzt, und auf den zusammenfassenden Bericht über eine Beratung zu Problemen der Mutationsinduktion im Erbgut extranuklearer Zellelemente wie Mitochondrien, Chloroplasten etc. Zahlreiche Eigenschaften verschiedenster Pflanzenarten erwiesen sich als vom Cytoplasma beeinflußt, z. B. Herbizidresistenz bei Brassica und Chenopodium und Blütenmorphologie bei Nicotiana. Es werden Empfehlungen gegeben, die Forschung auf diesem Gebiet zu intensivieren und die Ergebnisse dieser Untersuchungen verstärkt zu propagieren wegen der großen Bedeutung Cytoplasma-vererbter Merkmale bei Nutzpflanzen.

Für den Leser nützlich ist eine Tabelle über zahlreiche international übliche Einheiten und deren Umrechnung in SI-Einheiten am Schluß des Buches.

G. Hübner

Radiation Processing for Plastics and Rubber
Internat. Conference in Brighton/Sussex, 15. – 17. Juni 1981

London: The Plastics and Rubber Institute 1981; 312 S., Beilage 32 S., 144 Abb., 30 Tab.

Vor ca. 30 Jahren wurde begonnen, Bestrahlungstechnologien zur Plastmodifizierung industriell einzuführen. Gegenwärtig verkörpert die Industrieproduktion auf diesem Gebiet einen Wert von einer Mrd. \$, wobei die jährliche Wachstumsrate über 20% beträgt.

Vor diesem Hintergrund ist die oben zitierte Konferenz zur gegenseitigen Information von Anbietern und Nutzern kommerzieller Strahlenquellen (Elektronenbeschleuniger, ^{60}Co-γ-Anlagen) zu sehen. Organisiert unter Schirmherrschaft des Plastics and Rubber Institute von *A. Charlesby*, einem Senior auf dem Gebiet der Strahlenchemie von Polymeren, wurden nach einer allgemeinen Einführung in 33 15minütigen Vorträgen analytische (6), anlagentechnische (8), ökonomische (5) sowie anwendungs- und verfahrenstechnische (17) Probleme diskutiert.

Dabei konzentrierte sich das Interesse besonders auf den Problemkreis der strahlenchemischen Vernetzung von Polyethylen- bzw. Polyvinylchlorid-Isolationen von Kabeln der Starkstromtechnik und der Telekommunikation.

Die analytische Charakterisierung makroskopischer Eigenschaften vernetzter makromolekularer Stoffe und von Pfropfcopolymeren durch mechanische und chemische Methoden wird weitgehend beherrscht und zur Prozeßsteuerung eingesetzt (*B. J. Lyons*, Großbrit.). Durch Einführung gepulster Kernresonanztechniken (Protonenresonanz zur Spin-Spin-Relaxationsmessung, ^{13}C-Resonanz mit Kombination der magnischen-Winkel-Rotation und der Kreuzpolarisation) gelingt die Reduzierung der Linienverbreiterung im wesentlichen auf die chemische Verschiebung und damit der mikroskopische Nachweis von Verzweigungsstellen in unlöslichen und festen Proben (*A. Charlesby*, Shrivenham; *T. Yoshida* u. a. Sapporo, Japan).

Zum technischen Einsatz in der Plastmodifizierung kommen vorwiegend Elektronenbeschleuniger (0,3 ... 12 MeV), deren Typen von den Herstellern (Radiation Dynamics, High Voltage u. a.) ausführlich diskutiert wurden. Vorteile in bezug auf geringe Abschirmung und damit einfachere Verfahrensführung haben niederenergetische Elektronenbeschleuniger ($\leq$ 300 keV) mit linear angeordneten Bürstenkathoden zur großflächigen Oberflächenbehandlung von Folien und Textilien (*V. Läuppi*, Div. Energy Sciences Int., Genf; *P. Holl*, Polymer-Physik, Tübingen).

Neben den tragenden Themen wurden viele verschiedene Einzelanwendungen der Bestrahlungstechnik zur Polymermodifizierung vorgestellt wie z. B.

- die Pfropfcopolymerisation von Butylacrylat auf Schweinsleder (*K. Pietrucha* u. *J. Kroh*, Łodz, VR Polen),
- die Polyethylenvernetzung zu Schäumen (*A. B. Paterson*, BXL Plastics).
- die Erzeugung strahlengepfropfter Polymerhydrogele für Prothesenzwecke in der Medizin (*W. R. Watson* u. *R. C. Stone*, Oxford)

und die Erzeugung von Schrumpffolien als Verpackungsmaterialien (*W. G. Baid*, Grace Co., USA).

Der Konferenzband ist ein anregendes, naturgemäß aber nicht lückenloses (7 Vorträge nur als Kurzreferate) Informationsmaterial für Wissenschaftler und Techniker, die auf dem Gebiet der Polymermodifizierung und der Strahlentechnik arbeiten. *O. Brede*

Geology and Geochemistry of Manganese
Vol. 1: General Problems, Mineralogy, Geochemistry, Methods

Editors: I. M. Varentsov (UdSSR) and Gy. Grasselly (VR Ungarn). Budapest; Akadémiai Kiadó 1980. 463 S., div. Abb. u. Tab.; Leinen M 105.—

Anläßlich des 25. Internationalen Geologenkongresses 1976 in Sydney fand das 2. Internationale Symposium zur Geologie und Geochemie des Mangans statt, das von der Kommission der IAGOD und der Arbeitsgruppe des IGCP, Projekt Nr. 111 (Genese der Manganlagerstätten) ausgerichtet wurde. Die dort gehaltenen Vorträge werden nun in 3 Bänden in gedruckter Form vorgelegt.

Der 2. Band, ,,Manganese deposits on the continents" behandelt allgemein die Genese und am umfangreichsten einzelne Lagerstätten auf allen Kontinenten. Der 3. Band, ,,Manganese on the bottom of recent bassins", berichtet über Mn-Ausfällungen in marinen Tiefsee- und Flachwassergebieten sowie in einem gesonderten Kapitel übrr mikrobiologische Aspekte und in einem Beitrag über technische Gewinnungsmöglichkeiten im marinen Bereich.

Der hier näher vorzustellende 1. Band umfaßt 7 Beiträge zur Mineralogie, 5 Beiträge zur Geochemie und 2 zur Methodik.

Die umfangreichste Arbeit ist die über die Manganminerale von *G. Frenzel* mit 134 S., 66 Tab., 83 Abb. sowie 435 Literaturzitaten. Die Minerale werden einzeln behandelt und geben in gesonderten Abschnitten Auskunft über die Chemie, Kristallographie, Textur (Paragenese, Verwachsungsverhältnisse), physikalische Eigenschaften, Polierverhalten, Reflexionsvermögen (für unterschiedliche Wellenlängen), Ätzverhalten, thermochemische Eigenschaften, Infrarotspektren, Vorkommen, Diagnosemöglichkeit und Literatur. Zahlreiche Mikrophotographien von polierten Anschliffen, Abbildungen von dazugehörigen IR-Spektren und DTA-Kurven, Röntgendiagrammen (Diffraktometrie und *Debye-Scherrer*) sowie ein Tabellenanhang mit den *d*-Werten der Minerale lassen diesen Beitrag für diagnostische Arbeiten besonders wertvoll erscheinen.

Weitere Beiträge sind:

R. Giovanoli: Über natürliche und synthetische Manganknollen (vorwiegend kristallographische Untersuchungen mit Ableitungen zum Ausfällungsmechanismus);

R. K. Sorem u. *R. H. Fewkes*: Verteilung von Todorokit und Birnessit in Manganknollen in der Horn Region, E-Pazifik;

F. V. Cuchrov u. a.: Eisenoxide als Minerale in sedimentären Einheiten;

R. M. McKenzie: Manganoxide in Böden;

Tj. Peters, u. a.: Progressive Metamorphose von Mangankarbonaten und Cherts in den Alpen;

A. A. Sidorov u. a.: Manganminerale auf Gold-Silber-Lagerstätten.

Aus dem Kapitel ,,Geochemie" wären die Beiträge von *D. W. Crerar* u. a. und von *K.-H. Wedepohl* zur Übersicht über das geochemische Verhalten des Mangans in den einzelnen Geosphären hervorzuheben, die sich kaum überschneiden, sondern wechselseitig ergänzen.

Die weiteren Beiträge beschäftigen sich mit der Geochemie und Bildung von Mangandioxid-Lagerstätten (*Yu. Hariya*), mit der Geochemie der Übergangsmetalle (Mn, Ni, Co, Cu, V, Cr, Mo) bei der Bildung von Eisen-Manganerzen in marinen Becken (*I. M. Varentsov*) und mit der Bildung von metallorganischen Verbindungen des Mn und anderer Metalle im marinen Bereich, eine sehr moderne Arbeit.

Die beiden methodischen Arbeiten sind einmal der Analytik gewidmet — Emissionsspektralanalyse und Röntgenfluoreszenz (*J. C. Mills* u. a.), zum anderen der Anwendung statistischer Methoden bei der Aufklärung des Zusammenhanges von Sediment und Manganknollen im S-Pazifik (*M. Defossez* u. a.).

Der Verlag hat das Buch hervorragend ausgestattet, das betrifft Einband und Papier, Schriftsatz und besonders die Wiedergabequalität der mikrophotographischen und elektronenmikroskopischen Aufnahmen. Ein Sachwortverzeichnis am Schluß ermöglicht ein schnelles Auffinden von Sachverhalten in den unterschiedlichen Beiträgen.

Der Raum ist viel zu knapp, um die Einzelbeiträge einzuschätzen. Sie alle haben ein hohes wissenschaftliches Niveau, bestechen durch den komplexen Einsatz moderner und traditioneller (!) Meßmethoden, guten Darstellungen in Bildern und Diagrammen.

Dieser erste Teil der wohl bisher umfangreichsten Mangan-Monographie ist besonders Geochemikern, Mineralogrn, Kristallographen, Sedimentologen, Lagerstättenkundlern, Hydrogeologen, Bodenkundlern und im Umweltschutz tätigen Kollegen sowie auch Studenten dieser Disziplinen zu empfehlen. Um dieses Wissen kann man auf diesem Gebiet keinen Bogen machen. *P. Beuge*

Geochemistry of Coal

V. Bouška, Prague: Academia 1981. 284 S., 72 Abb., 17 Fotos, 44 Tab.; geb. M 46,80

Es ist ein anspruchsvolles, fast ehrgeiziges Ziel, eine Geochemie der Kohle zu schreiben. Der Autor, Dozent an der

Karls-Universität in Prag, hat dieses Buch nicht nur für Geochemiker, sondern auch für Geologen und Techniker, die mit dem Rohstoff Kohle zu tun haben, geschrieben. Die 1. Auflage des Buches in tschechischer Sprache war kurz nach Erscheinen 1975 bereits vergriffen, und so erschien bald darauf die 2. Auflage (1981), allerdings in englischer Sprache, die nunmehr vorliegt.

Das Grundkonzept des Buches besteht aus folgenden Teilen: chemischer Aufbau und chemische Zusammensetzung von Pflanzen, Vergleich der chemischen Zusammensetzung von fossilen und rezenten Pflanzen und die chemische Zusammensetzung der Kohle und der Asche. Dieser hervorgehobene geochemische Aspekt wird durch die geologischen Veränderungen des kohlebildenden Pflanzenmaterials (Torfstadium, Inkohlung) und durch die Mineralogie des Torfes und der Kohle flankiert. Die Mechanismen der Anreicherung und Verteilung von Spurenelementen in Kohle und damit in Zusammenhang stehender Fragen der Stratigraphie von Kohleflözen dienen der Abrundung des Themas.

Bei der alphabetischen Aufzählung fast aller Elemente des Periodensystems wird nicht deutlich, welche Elemente in echter Weise am Stoffwechsel der Pflanzen teilnehmen, eben Bioelemente sind, und welche offensichtlich durch Stoffströme aus dem Boden (oder Aufnahme aus der Luft) in das Pflanzenmaterial gelangt sind. Gleiches gilt auch von dem Stoffbestand der Kohle, wobei durch Migrationsprozesse in der Bodenzone die Vielfalt an Elementen, die sich in der Kohle (Aschen) befinden, lediglich zunimmt. Außerdem würde eine tabellarische Zusammenstellung völlig ausreichend sein.

Der organische Stoffbestand ist sporadisch und keineswegs treffend dargestellt. Beispielsweise werden die Arbeiten zum Lignin von *K. Freudenberg*, *K. Lundquist*, *H. Nimz*, die Standardwerke von *F. E. Brauns*, *I. A. Pearl* u. a. gar nicht erwähnt und moderne Analysenverfahren der Strukturaufklärung (^{13}C-NMR-Technik) nicht berührt. Die Isotopenvariationen des Kohlenstoffs und Stickstoffs lassen nicht erkennen, welcher Zusammenhang zwischen $\delta^{13}C$-Werten (und nicht Verhältnissen, wie immer gesagt wird) und dem Inkohlungsgrad bzw. dem Reflexionsvermögen am Vitrinit besteht. Auf S. 130 wird ein Standard eingeführt, der offensichtlich durch eine falsche Transliteration zustande gekommen und nur schwierig wiederzuerkennen ist: Es wird von einem RDV-Standard gesprochen, wobei PDB (**Pe**edee-Formation eines **B**elemniten aus Südkarolina — durch *H. Craig* 1957 eingeführt) gemeint ist. Der Luft-Standard des Stickstoffs wird auf S. 179 gar nicht erwähnt. Hier bedarf es mehr als nur der Mitteilung von einzelnen Zahlenwerten. Von der isotopen Zusammensetzung des Schwefels in Kohle wird auf S. 182 nur gesagt, daß sie von *Smith* und *Batts* (1974) studiert worden wäre — und die mögliche SO_2-Zuordnung zu bestimmten Emittenten wird beispielsweise auch nicht deutlich. Von den rein geologischen Angaben wäre zu erwähnen, daß auf S. 13 präkambrische Kohlevorkommen in Kasachstan angegeben werden, wo doch die Pflanzen das Land erst im Silur/Devon erobert haben. Wie der Urangehalt in Kohle zustande kommt, der u. a. für die radioaktiven Emissionen in Kohlekraftwerken verantwortlich gemacht wird, bleibt auch offen (Na-uranyl-dicarbonat und -tricarbonat). Die Literatur ist mit insgesamt 696 Zitaten erfaßt, allerdings hauptsächlich mit älteren Arbeiten; so beträgt der Anteil der Zitate aus den Jahren 1971—1979 nur 18,4%. Die Bedeutung der Kohle als Energieträger und ihr steigender Einsatz in den kommenden Jahren ist auch nicht befriedigend dargestellt; eine einzige Weltproduktionsangabe aus dem Jahre 1974 reicht dazu nicht aus, abgesehen von Problemen der Kohle-Vergasung (Hochtemperaturreaktor) und den Umweltbelastungen durch die Kohleverbrennung (saurer Regen) u. ä. m.

Im Vorwort zur 2. Auflage wird dieses Buch als ein Handbuch bezeichnet. Dieser Anspruch ist zu hoch. Es bedarf einer grundsätzlichen Aktualisierung und vor allem Akzentuierung, um diesen Anspruch, ein Handbuch zu sein, erfüllen zu können.

H. Hübner

Environmental Isotopes in the Hydrosphere

V. J. Ferronski u. V. A. Polyakov, übersetzt von S. V. Ferronski. Chichester — New York — Brisbane — Toronto — Singapore; John Wiley & *Sons 1982, 466 S., 133 Abb., 87 Tab., geb. £ 31,00.*

Im Jahre 1975 haben *V. J. Ferronski* et al. eine Monographie mit dem Titel „Natürliche Isotope der Hydrosphäre“ in russischer Sprache publiziert. Diese Monographie liegt nun in englischer Übersetzung vor, allerdings in einer stark veränderten und aktualisierten Form. Die Veränderung betrifft den Verzicht, meßtechnische Einzelheiten zur Bestimmung von Isotopenverhältnissen und niedrigen Aktivitäten darzustellen; die Aktualisierung ist vor allem in Hinsicht auf die in der Zwischenzeit veröffentlichte Literatur zu sehen, denn allein 24,6% der angegebenen Literatur entfallen auf die Jahre 1976—82 mit einem Maximum von rund 10% aus dem Jahr 1979, wobei insgesamt 796 Literaturstellen aufgeführt sind.

Das Buch ist in 5 große Kapitel gegliedert: Isotopengeochemie der natürlichen Wässer (S. 7—80), stabile Isotope des Wasserstoffs und Sauerstoffs beim Stadium der Bildung und der Dynamik der natürlichen Wässer (S. 81 bis 202), kosmogene radioaktive Isotope in den natürlichen Wässern (S. 203—323), radioaktive Isotope der schweren Elemente im Wasserkreislauf (S. 324—370) und ein fünftes und letztes Kapitel, das eine Aussage über den Ursprung der Hydrosphäre der Erde im Lichte der isotopen Zusammensetzung des Wassers trifft (S. 371—418).

Im 1. Kapitel wird auch die Theorie der Isotopieeffekte behandelt. Hier sollte manches dem englischen Sprachraum mehr angepaßt sein. So werden die β-Faktoren nach *Varshavsky* und *Veisberg* (1955) eingeführt, ohne ihre Entsprechung mit dem reduzierten Zustandssummenverhältnis nach *Bigleisen* und *Mayer* (1947) deutlich zu machen. Ferner sollte an einer Stelle gesagt werden, daß ohne Berücksichtigung der Anharmonizität bei den Wasserstoffisotopen eine befriedigende Übereinstimmung zwischen Theorie und Experiment nicht erzielbar ist; hier dürften in dem molekularen Kraftfeldansatz über die quadratischen Glieder hinaus auch kubische Glieder und Glieder 4. Grades für eine Auslenkung unbedingt erforderlich sein. Bei den sowjetischen Autoren traten auch mitunter uneinheitliche Transkriptionsprobleme auf: *Varshavsky* und *Warshavsky* auf S. 20 oder *Vinograd* und *Winograd* S. 123, 124, 457; auch Rücktranskriptionen können Schwierigkeiten bereiten, S. 7 und S. 400 *Clapeiron* für *Benoît Pierre Emile Clapeyron*, den französischen Thermodynamiker, der zum erstenmal den *Carnot*schen Kreisprozeß in eine mathematische Form brachte. Daß es für viele Leser ungewohnt sein dürfte, die bekannte *Clausius-Clapeyron*-Gleichung als *Clapeyron-Clausius*-Gleichung benannt zu bekommen, sei nur am Rande erwähnt. Bei dem Wasserstandard SMOW fehlt ein Hinweis auf den seit einigen Jahren substantiell wieder vorhandenen IAEA-Standard, dem Vienna-SMOW oder kurz V-SMOW, der mit dem Original-SMOW nahezu übereinstimmt. Ein SLAP-Standard (Standard Light Antarctic Precipitation) ist insofern erwähnenswert, als man damit bei internationalen Vergleichen der Standards (Intercomparisms) eine gute Skala zur Verfügung hat. Die auf S. 187 erläuterte Reaktion des CO_2/CH_4-Systems in der Geothermie von hydrothermalen Wässern ist keine *Fischer-Tropsch*-Reaktion, die ja zwischen CO und H_2 in Anwesenheit von Katalysatoren und bei 200–400 °C unter Bildung höherer Kohlenwasserstoffe und Alkohole abläuft, so wie sie auf S. 389 angegeben wird. Inwieweit diese Reaktion als Quelle abiogener organischer Verbindungen in einer Protoerde eine Rolle gespielt hat, ist eine Frage der Bedingungen, beispielsweise in Anwesenheit von NH_3 oder nicht, da sie Aminosäuren, Purine, Pyrimidine u. a. m. bilden können oder nicht. Der auf S. 422 aufgeführte Autor heißt *Bigl, F.* und nicht *Bige*, cherts und nicht chers S. 149 und *Gaussian* S. 269. Bei allen aufgeführten Tabellen und Übersichten erfahren die Stickstoffisotope $^{14}N/^{15}N$ keinerlei Erwähnung, obwohl sie durchaus mit zu den Umweltisotopen zu rechnen sind und in landwirtschaftlich stark genutzten Arealen in den Grundwässern eine wachsende Bedeutung haben. Daß dabei biologische und pedosphärische Prozesse eine Rolle spielen, erschwert zwar ihre Interpretation im Detail, begründet aber nicht ihre Negation; außerdem ist eine zunehmende Bearbeitung in der Literatur beobachtbar.

Die Monographie stellt eine umfassende Zusammenstellung der Isotopenmethoden in der Hydrosphäre dar, wobei auch erstmals die stärkere Berücksichtigung der sowjetischen Literatur auf diesem Gebiet bemerkenswert ist. Die Einbeziehung von Prozessen aus geologischen Zeiten bis hin zur Entstehungsgeschichte unseres Planeten ist ein weiterer Vorzug dieses Buches und erweitert damit den Interessentenkreis sehr wesentlich. Eine Leserschaft von Geochemikern, Hydrologen und von Wissenschaftlern, die von Seiten der Isotopenmethoden diese Problemkreise flankieren, dürfte oft diese Monographie zur Hand nehmen, um verbindliches Wissen auf diesem Spezialgebiet nachzuschlagen.

H. Hübner

Industrial Application of Radioisotopes and Radiation Technology
Conference Proceedings, Grenoble 28. Sept.–2. Oct. 1981

Vienna: IAEA 1982, 594 S., zahlr. Tab. und Abb.; karton. Austrian Schillings 980,–

An der Konferenz beteiligten sich 278 Teilnehmer aus 39 Ländern und zwei internationale Organisationen. Der vorliegende Band der Proceedingsreihe der IAEA ist der Konferenzbericht der in 8 Sektionen gebotenen 19 Plenar- und 39 Postervorträge, der Forumsdiskussionen zu den Themen:

- Strahlenbehandlung von Abfällen und deren Wiedernutzung,
- Fortschritte in der industriellen Strahlentechnik,
- Weiterentwicklungen in der Gerätetechnik,
- Neues auf dem Gebiet der Bestrahlungsanlagen,
- Ökonomische Ergebnisse der Isotopen- und Strahlentechnik und des Rundtischgespräches über die Einführung der Isotopen- und Strahlentechnik in Entwicklungsländern.

Die Vorträge sind zu folgenden Schwerpunktthemen (Sektionen) zusammengefaßt:

- Kerntechnik bei der Förderung, Verarbeitung und Rückgewinnung von Rohstoffen,
- Strahlentechnik in der Industrie,
- Kerntechnik im Umweltschutz,
- Strahlenbehandlung in der Biotechnik,
- Isotopentechnik bei der Betriebs- und Qualitätskontrolle in der Industrie,
- Kerntechnische Geräte und Bestrahlungs-Anlagen,
- Technologietranfer in Entwicklungsländer und
- Ökonomischer Nutzen der Kerntechnik.

In dem Band sind die Plenarvorträge vollständig, die Poster in Kurzfassung enthalten. Die Diskussionen zu Vorträgen sind am Ende einer Vortragsreihe zusammenfassend beschrieben. Die Diskussionen der Foren und des Rundtischgespräches werden ebenfalls, meist recht informativ, ausführlich und mit Zahlenmaterial belegt, behandelt. Einige interessante in den Proceedings beschriebene Aspekte der Kerntechnik sollen erwähnt werden, um den breiten Interessentenkreis für diesen Band anzudeuten. Bei der Prospektierung von Erzen wird die Neutronentechnik als bevorzugte Methode beschrieben; für die Technik muß jedoch ein preisgünstiger Hochleistungs-Neutronengenerator im Angebot sein. Weitgehend automatisierte Geräte, die nach dem Prinzip der Anregung durch prompte γ-Strahlung mittels Pu/Be-Neutronenquellen arbeiten, werden für eine weitgehende Kohleanalytik als aussichtsreich angesehen.

Die Nutzung strahlenchemischer Verfahren in der Industrie entwickelt sich nur langsam; sie wird trotz teilweise unbestreitbarer Vorteile mit Skepsis betrachtet. Dagegen haben sich Technologien mit Elektronenbeschleunigern zur Plastmodifizierung weltweit durchgesetzt. Zu Fragen der Strahlenbehandlung von Abfällen ist man der Meinung, daß Abwässer von Großstädten mit Niedrigenergie-Beschleunigern und solche von kleineren Städten vorteilhaft mit Radionuklidquellen oder Hochenergie-Beschleunigern behandelt werden sollten, um dann einer Wiedernutzung zugeführt zu werden. Der Umfang der Strahlensterilisation von medizinischen Artikeln und Produkten steigt weiter an, Implantate hoher Widerstandsfähigkeit und Bioverträglichkeit lassen sich durch Strahlenbehandlung herstellen. Im Umweltschutz werden strahlen- und isotopentechnische Methoden verstärkt Eingang finden, wie z. B. Verfahren der Ionisationsgasanalyse zum Nachweis toxischer Stoffe. Aus den USA werden andererseits erste Ergebnisse über eine strahlenchemische Rauchgasreinigungsanlage berichtet.

Das weiteste Feld der Anwendung von Radionukliden ist die Isotopen- und Strahlentechnik; sie ist zur Standardtechnik geworden. Es werden zwar nur wenige prinzipiell neue Anwendungsmöglichkeiten geboten, dafür wird aber ein immer breiteres Anwendungsfeld erschlossen. In notwendiger Ergänzung konventioneller Methoden wird die Tracertechnik zur Prozeßanalyse genutzt. Bei der Entwicklung kerntechnischer Geräte bringt der Einsatz von Mikroprozessoren einen entscheidenden Fortschritt; dabei ist aber die Entwicklung von betriebssicheren, selektiven und empfindlichen Detektoren nicht zu vernachlässigen. Interessante und detaillierte Angaben enthält der Konferenzbericht zum Nutzen der Isotopen- und Strahlentechnik.

Der Band enthält am Schluß eine kurze Zusammenfassung der Ergebnisse der Konferenz, eine Autorenliste und eine Umrechnungstabelle für Maßeinheiten in das Internationale Einheitensytem. Mit der Herausgabe des Bandes soll die Nutzung der Methoden noch weiter initiiert werden. Die einzelnen Beiträge sind nicht redaktionell bearbeitet; sie unterscheiden sich deshalb in ihrem Informationsgehalt. Da der Konferenzbericht neben teilweise sehr interessanten Anwendungsbeispielen Tendenzen in der Strahlen- und Isotopentechnik aufzeigt, ist er einmal allen auf diesem Gebiet tätigen Wissenschaftlern und Technikern sehr zu empfehlen, zum anderen zeigt er Fachleuten der Industrie zahlreiche sehr aktuelle Anwendungsmöglichkeiten.

R. Otto

Tabellen der Spektrallinien von Atomen und Ionen Handbuch

A. R. Striganow, G. A. Odinzowa, Moskau: Energoizdat 1982. 312 S., 109 Tab., 253 Lit. (in Russ.); M 14,—

Seit der Herausgabe des bekannten Vorläufers dieses Handbuches durch *A. R. Striganov* und *N. S. Sventitskij* sind 18 Jahre vergangen, in denen sich die Zahl der vermessenen Atom- und Ionenlinien beträchtlich erhöht und die Genauigkeit der Angaben wesentlich verbessert haben.

Etwa analog der älteren Ausgabe enthält dieses Handbuch Spektrallinien der ersten 18 Elemente des Periodensystems, vom Wasserstoff bis zum Argon. Indem es vom Helium bis zum Neon alle Ionen einschließlich der wasserstoffähnlichen, bis zum Schwefel die 1—5fach geladenen und vom Argon auch noch die 9fach geladenen beinhaltet, ist es die derzeit umfassendste Sammlung von Spektrallinien für diese Elemente.

Die erfaßten Wellenlängen erstrecken sich vom Vakuum-Ultraviolett bis zum Infrarot (etwa 1 — 40000 nm). In bewährter Weise werden die 30490 Spektrallinien, geordnet nach Elementen und Ionisierungszuständen, durch Angabe der Wellenlänge in *Ångström*-Einheiten, eines Relativwertes der Linienintensität, der beiden Niveauenergien in eV, des Elektronenübergangs und der beiden Gesamtdrehimpuls-Quantenzahlen J tabelliert. Für jedes Element wird eine kurze Einführung mit einer Charakterisierung und Wertung der verwendeten Quellen gegeben.

Die Erweiterung gegenüber dem ersten Handbuch besteht vor allem in der Aufnahme neuer Interkombinations-, Autoionisierungs- und Zweielektronenübergänge, die durch die enormen Fortschritte auf dem Gebiet der Quantenelektronik gefunden wurden, und in der dadurch möglichen Vervollständigung der Energieniveausysteme dieser Elemente. Die Verbesserung besteht in der Revision der Daten durch Berücksichtigung genauerer Messungen und in der Korrektur zahlreicher Fehler in der Niveauzuordnung. Als umfassendes Tabellenwerk dient die vorliegende Ausgabe insbesondere dem Physiker und dem Ingenieur, der sich mit Plasmauntersuchungen, elektrischen Entladungen und Atomspektroskopie beschäftigt oder auf dem Gebiet der Quantenelektronik und Astrophysik arbeitet.

G. Müller

Mineralogische Tabellen

H. Strunz, C. Tennyson. 8. Aufl.; Leipzig: Akad. Verl.-Gesellsch. Geest & Portig 1982; M 47,— (Ausland: 54,—)

In nunmehr 8. Auflage sind die „Mineralogischen Tabellen" von *Hugo Strunz* unter Mitarbeit von *Chr. Tennyson* erschienen. Damit ist ein Kompendium für Mineralien wieder verfügbar geworden, das sich seit seinem ersten Erscheinen 1941 nicht nur einen festen Platz in den Fachbibliotheken und den Bücherregalen der Geowissenschaftler und Liebhabermineralogen erobert hat, sondern vor allem als Standard- und Nachschlagewerk einzig dastehend unentbehrlich geworden ist. Die „Strunzsche Mineralsystematik" hat sich weltweit Anerkennung und Geltung verschafft und bildet heute grundlegendes Ordnungsprinzip jeder wissenschaftlichen Mineralsammlung. Gegenüber den ersten Auflagen mehr als verdoppelt, vereinigt es nach strenger kritischer Auswahl alle ein Mineral definierenden chemischen und strukturellen Daten für 1745 eigenständige Spezies, die bis 1969 ermittelt worden sind.

Ein erster Teil — Einführung in die Kristallchemie — ist in zwei Abschnitte gegliedert. Im ersten findet der Leser auf 30 Seiten komprimiert die Grundlagen der Kristallchemie (Definitionen, Grundgesetze des chemischen und strukturellen Aufbaues der Kristalle), die Klassifikationsprinzipien der Mineralien sowie in Tabellenform die wichtigsten allgemein gültigen Daten (Kristallklassen, Raumgruppen, Periodensystem der Elemente, Größen der Kristallbausteine). Der zweite Abschnitt beinhaltet einen in dieser Ausführlichkeit in keinem anderen Buch anzutreffenden Überblick über sämtliche häufigeren Kristallstrukturtypen mit Hinweisen auf isotype und homöotype Realisierungen.

Im zweiten Teil des Buches — mit 408 Seiten den Schwerpunkt bildend — werden über 1700 Mineralarten auf krirtallchemischer Grundlage geordnet mit internationaler Bezeichnung (mit Erstautor und Jahresangabe), chemischer Formel, Kristallklasse und Raumgruppe, Gitterkonstanten und kristallographischem Achsenverhältnis, Zahl der Formeleinheiten pro Elementarzelle und Strukturtyp aufgeführt. Von besonderem Wert sind die in zahlreichen Fußnoten gegebenen Hinweise auf die jeweiligen Originalarbeiten. Auch 400 eingebürgerte Varietätennamen sind bei den entsprechenden Mineralien genannt. Besonders dieser Teil des Buches ist als Unikat in der modernen mineralogischen Literatur von erstrangigem wissenschaftlichem Wert.

Der dritte Teil des Werkes umfaßt die Register: Namenregister zur Systematik und ausgeschiedene Mineralnamen sowie Formelregister zur Systematik der Mineralien mit einem Umfang von 91 Seiten zeigen, daß hier nicht nur ein Anhang, sondern ein wichtiges wissenschaftliches Arbeitsmittel vorliegt. Die etwa 7800 Stichwörter beinhalten Mineralnamen, die zur Systematik gehören, veraltete und daher nicht mehr anzuwendende Bezeichnungen, ungenau definierte „Minerale", Synonyma, Varietäten, Pseudomorphosenbezeichnungen und Handelsnamen.

Die Form der Darstellung, Satz und Druck entsprechen der Bedeutung des Werkes und sind vorbildlich. Es kann allen Geowissenschaftlern, Laienmineralogen, aber auch allen mit anorganischen Naturprodukten und Werkstoffen Befaßten sehr empfohlen werden, dieses Werk zu erwerben. Das gilt auch für diejenigen, die bereits frühere Auflagen besitzen, da die vorliegende wesentlich erweitert und aktualisiert ist. Autoren und Verlag ist für diese mit größter Sorgfalt durchgeführte Arbeit besonders zu danken.

P. Schreiter

Mineral Deposits and the Evolution of the Biosphere
Physical and Chemical Sciences Research Report 3

H. D. Holland and M. Schidlowski (ed.). Berlin — Heidelberg — New York: Springer 1982. 333 S., 40 Abb., 10 Tab., 4 Fotos (Dahlem Workshop Reports); geb. DM 42,— ; US $ 19.60

Vom 1. — 5. September 1980 fanden sich 46 Wissenschaftler in Berlin-Dahlem zusammen, um gemeinsam über mineralische Ablagerungen und ihren Zusammenhang mit der Evolution der Biopshäre zu diskutieren. Solche und ähnliche Zusammenkünfte im Rahmen einer Arbeitstagung wären an sich nichts ungewöhnliches, aber die seit 1974 kreierten Dahlem-Konferenzen bekunden schon etwas besonderes: Es werden Themen behandelt, die von hohem internationalem Interesse, von besonderer interdisziplinärer Verflochtenheit, die aktuell und ausgesprochen problemorientiert sind. Die Teilnehmer sind auf die Zahl 48 beschränkt und werden durch einen beratenden Ausschuß (Problem Advisory Committee) ausgewählt. Das Besondere an dem Dahlem-Konferenz-Modell ist noch die Tatsache, daß keine Vorträge gehalten werden, sondern daß in 4 Untergruppen zu je 12 Mitgliedern die eigentlichen Diskussionen auf der Grundlage von vorher ausgearbeiteten Arbeitspapieren (background papers) stattfinden. Das wesentliche Ergebnis dieser Arbeitstagung sind die 4 Statusberichte, die von Berichterstattern (Rapporteurs) vorgetragen werden und die versuchen, den derzeitigen Stand unseres Wissens in kurzer Form darzulegen.

Eine Arbeitsgruppe behandelte die sedimentären Eisenlagerstätten, die Evaporite und Phosphorite. Die Entstehungsgeschichte der riesigen Eisenlagerstätten aus dem Präkambrium (**b**anded **i**ron **f**ormation = BIF) war wieder Gegenstand eines regen Meinungsaustausches, und unter anderem konnte als Ergebnis festgehalten werden, daß die Mikrobandstrukturen nicht notwendigerweise nur durch eine sedimentäre Ablagerung von Fe-III, sondern auch durch eine bakterielle Reduktion im Eisen, seine Auflösung und sein Abtransport aus der chert-Eisenoxidmischung möglich ist. Die zweite Arbeitsgruppe behandelte die Schwefellagerstätten, die durch bakterielle Reduktion von Sulfat in Sulfid entstanden sind. Hier wurden besonders die $\delta^{34}S$-Variationen diskutiert, die trotz umfangreicher weltweiter Untersuchungen immer noch manche Frage offen lassen, beispielsweise folgende: Warum beobachtet man unter Laborbedingungen maximale Isotopenfraktionierungen des Schwefels von ungefähr 35‰, während in der Natur durchaus Werte von rund 60‰ erreicht werden können? Die dritte Arbeitsgruppe behandelte den reduzierten organischen Kohlenstoff in Sedimenten und berührte damit viele Fragen der Evolution der Biosphäre. Hierzu ist eigentlich das gesamte Arsenal an moderner Instrumentalanalytik gefordert: hochauflösende Massenspektrometrie (MS), Kopplung GC mit MS, ^{13}C-NMR, Laser-Pyrolyse- und Laser-Ionisations-MS, Bestimmung der ^{13}C-Variationen innerhalb des Kohlenstoffgerüstes von organischen Verbindungen u. a. m. Von den $\delta^{13}C$-Werten dieser Stoffgruppe, die älter als $3{,}5 \cdot 10^9$ a sind, kann nur gesagt werden, daß autotrophe Mikroorganismen für ihr Entstehen verantwortlich gemacht werden können, nichts aber über den Kohlenstoff-Metabolismus im Detail. Eingehend wurden auch die Werte der Isua-Sedimente diskutiert, wobei die graphitischen Kerogenreste für einen eindeutigen Beweis für die Gegenwart von Leben nicht ausreichend zu sein scheinen. Die vierte Arbeitsgruppe behandelt die biogeochemische Evolution des Ozean-Atmosphäre-Systems, wobei insbesondere die geologische Entwicklung des Sauerstoffs in der Atmosphäre eingehend behandelt wurde.

Nicht nur die behandelten Themen waren höchst aktuell, sondern auch die dazu angegebene Literatur, die durch insgesamt 525 Zitate belegt wurde, wovon allein rund 63% aus den Jahren 1975—1982 stammen. Eine Literaturstelle (die 28 auf S. 106) ist im Verzeichnis S. 120/121 flüchtigerweise nicht mit angegeben. Die Arbeit von *E. C. Perry* und *F. C. Tan* 1972 oder die von *B. Nagy* et al. 1974 könnte gemeint sein.

Hier liegt eine aus berufenem Munde ausgesprochene Zusammenfassung des gegenwärtigen Wissensstandes vor, die auch richtungweisend für weiter zu bearbeitende Teilaspekte ist; ein bedeutender Beitrag zur Evolution der Biosphäre, die wohl in künftigen Veröffentlichungen sehr oft zitiert werden wird.

H. Hübner

Das Schicksal der Erde
Gefahr und Folgen eines Atomkriegs

J. Schell. München: R. Piper & Co. 1982. 267 S. (aus dem Amerikan.); kart. DM 19,80

Selten habe ich ein Buch mit so viel Interesse, aber auch mit so zwiespältigen Gefühlen gelesen. Lange habe ich gezögert, ehe ich mich dazu entschlossen habe, der Aufforderung zu folgen, es zu rezensieren, weil ich dem Autor in seinen Vorstellungen über die Folgen eines Kernwaffenkrieges in wenigstens einem wichtigen Punkte nicht und in seinen weltanschaulichen Grundpositionen überhaupt nicht folgen kann.

Im ersten Abschnitt werden die Wirkungen von Kernwaffen und die Folgen eines Kernwaffenkrieges beschrieben. Es wird dargelegt, daß ein globaler Kernwaffenkrieg gräßliche Folgen für die Menschen in den betroffenen Ländern, aber auch in anderen Ländern hätte, was zweifellos zutrifft. Eine entscheidende Schwäche des Buches besteht jedoch darin, daß davon ausgegangen wird, daß ein solcher Krieg zu einer vollständigen Vernichtung der Menschheit und vieler Pflanzen- und Tierarten führen könnte bzw. führen würde (s. z. B. S. 215 u. f.). An dieser Stelle werden neben den unmittelbaren Kernwaffenwirkungen auch die Folgen des Abbaus des Ozonschilds durch auf strahlenchemischem Wege in der Atmosphäre gebildete nitrose Gase ins Feld geführt. Auch die Annahme, daß alle Überlebenden, sofern es solche gibt, unter dem Einfluß der Sofortstrahlung und des Fallout steril sein könnten (S. 191), hält einer Prüfung nicht stand. Die Auswirkungen eines globalen Kernwaffenkrieges auf von den von diesem Krieg betroffenen Ländern entfernte Gebiete wird überschätzt. Hinzu kommt, daß bei der Darstellung der physikalischen Grundlagen Fehler unterlaufen sind, die sicherlich nur zu einem Teil dem Übersetzer angelastet werden dürfen. (Wer sich über die Auswirkungen eines Kernwaffenkrieges informieren will, sei auf AMBIO XI, Heft 2/3 (1982) verwiesen.)

Im zweiten Abschnitt werden die sich aus der Existenz der Kernwaffen ableitenden weltpolitischen Konsequenzen dargelegt. Man kann dem Autor zustimmen, wenn er darlegt, daß ein Krieg zwischen Kernwaffen besitzenden Ländern rasch zu einem globalen Kernwaffenkrieg mit seinen grauenhaften Konsequenzen eskalieren würde, daß die Konzeption (der Rezensent fügt hinzu: die Konzeption der imperialistischen Länder) von der Sicherheit durch nukleare Abschreckung absurd ist und daß die Menschheit dem

Risiko einer nuklearen Katastrophe nur durch Beendigung des Wettrüstens und durch Abrüstung entgehen kann. Leider übersieht der Autor in diesem Abschnitt die unterschiedlichen Positionen, die die UdSSR und andere sozialistische Länder auf der einen Seite und imperialistische Länder wie die USA auf der anderen Seite zu diesen Fragen einnehmen.

Im dritten Abschnitt des Buches versucht der Autor, den Ausweg aus der durch die Existenz der Kernwaffen heraufbeschworenen politischen Lage zu zeigen: Aufgabe der nationalen Souveränität und (dadurch möglich werdender) Verzicht auf nukleare und konventionelle Waffen. ,,Unser gegenwärtiges System und seine Indikationen sind die Trümmer der Geschichte", heißt es auf S. 247. Ist damit das kapitalistische System gemeint? ,,Genauso wie wir beschlossen haben, in dem System souveräner Staaten zu leben, können wir beschließen, in irgendeinem anderen System zu leben" (S. 255). Wenn wir bedenken, daß Nationen mit ihrem Bedürfnis nach nationaler Souveränität im Schoße der kapitalistischen Gesellschaftsordnung entstanden sind, so liegt der Schluß nahe, daß mit diesem anderen System ein kommunistisches Weltsystem gemeint sein könnte. Darin würde ich dem Autor aus tiefer Überzeugung zustimmen, wenn ich auch nicht verstehen würde, weshalb er es dann für so folgenschwer hält, ,,daß es den Vereinigten Staaten nicht gelang, sich ein Kernwaffenmonopol zu verschaffen und die Sowjetunion daran zu hindern, Atomwaffen zu bauen" (S. 126).

K. Wetzel

Das Neutron

Eine Artikelsammlung

Herausgegeben von B. M. KEDROW

(Übersetzung aus dem Russischen und aus englischen und französischen Originalarbeiten)

In deutscher Sprache herausgegeben von ROLAND REIF

(Wissenschaftliche Taschenbücher, Reihe Texte u. Studien)

1979. 269 Seiten — 11 Abbildungen — kl. 8° — 12,50 M

Bestell-Nr. 7623862 (7214)

Der Leser erhält ein lebendiges Bild der historischen Entwicklung, die zur Entdeckung des Neutrons und als Folgeerscheinung zu einem bedeutenden Aufschwung der experimentellen und theoretischen Kernphysik führte. Im ersten Teil schildern namhafte Wissenschaftler, deren Arbeiten diesen Erkenntnisfortschritt mit bewirkten, die entscheidenden Experimente zum Nachweis des Neutrons und zur Entdeckung von künstlicher Radioaktivität, Kernspaltung und transuraner Elemente. Der zweite Teil behandelt wissenschaftshistorische und methodologische Aspekte der Entdeckungsgeschichte des Neutrons. Der Sammelband wird ergänzt durch einen Anhang, der die wichtigsten, heute zum Teil nur schwer zugänglichen Publikationen zum behandelten Problemkreis enthält.

Bitte richten Sie Ihre Bestellungen an eine Buchhandlung.

AKADEMIE-VERLAG

DDR-1086 Berlin, Leipziger Str. 3—4

Contents of this Issue

Indexed in Current Contents/PC & ES
Science Citation Index
and INIS Atomindex

FSC
www.fsc.org
MIX
Papier | Fördert gute Waldnutzung
FSC® C083411

Zeitfracht Medien GmbH
Ferdinand-Jühlke-Straße 7
99095 Erfurt, Deutschland
produktsicherheit@kolibri360.de